Compressed Sensing for Engineers

Devices, Circuits, and Systems

Series Editor

Krzysztof Iniewski
Emerging Technologies CMOS Inc.
Vancouver, British Columbia, Canada

PUBLISHED TITLES:

3D Integration in VLSI Circuits: Implementation Technologies and Applications
Katsuyuki Sakuma

Advances in Imaging and Sensing
Shuo Tang and Daryoosh Saeedkia

Analog Electronics for Radiation Detection
Renato Turchetta

Atomic Nanoscale Technology in the Nuclear Industry
Taeho Woo

Biological and Medical Sensor Technologies
Krzysztof Iniewski

Biomaterials and Immune Response: Complications, Mechanisms and Immunomodulation
Nihal Engin Vrana

Building Sensor Networks: From Design to Applications
Ioanis Nikolaidis and Krzysztof Iniewski

Cell and Material Interface: Advances in Tissue Engineering, Biosensor, Implant, and Imaging Technologies
Nihal Engin Vrana

Circuits and Systems for Security and Privacy
Farhana Sheikh and Leonel Sousa

Circuits at the Nanoscale: Communications, Imaging, and Sensing
Krzysztof Iniewski

CMOS: Front-End Electronics for Radiation Sensors
Angelo Rivetti

CMOS Time-Mode Circuits and Systems: Fundamentals and Applications
Fei Yuan

Design of 3D Integrated Circuits and Systems
Rohit Sharma

PUBLISHED TITLES:

Diagnostic Devices with Microfluidics
Francesco Piraino and Šeila Selimović

Electrical Solitons: Theory, Design, and Applications
David Ricketts and Donhee Ham

Electronics for Radiation Detection
Krzysztof Iniewski

Electrostatic Discharge Protection: Advances and Applications
Juin J. Liou

Embedded and Networking Systems: Design, Software, and Implementation
Gul N. Khan and Krzysztof Iniewski

Energy Harvesting with Functional Materials and Microsystems
Madhu Bhaskaran, Sharath Sriram, and Krzysztof Iniewski

Gallium Nitride (GaN): Physics, Devices, and Technology
Farid Medjdoub

Graphene, Carbon Nanotubes, and Nanostuctures: Techniques and Applications
James E. Morris and Krzysztof Iniewski

High-Speed and Lower Power Technologies: Electronics and Photonics
Jung Han Choi and Krzysztof Iniewski

High-Speed Devices and Circuits with THz Applications
Jung Han Choi

High-Speed Photonics Interconnects
Lukas Chrostowski and Krzysztof Iniewski

High Frequency Communication and Sensing: Traveling-Wave Techniques
Ahmet Tekin and Ahmed Emira

High Performance CMOS Range Imaging: Device Technology and Systems Considerations
Andreas Süss

Integrated Microsystems: Electronics, Photonics, and Biotechnology
Krzysztof Iniewski

Integrated Power Devices and TCAD Simulation
Yue Fu, Zhanming Li, Wai Tung Ng, and Johnny K.O. Sin

Internet Networks: Wired, Wireless, and Optical Technologies
Krzysztof Iniewski

Introduction to Smart eHealth and eCare Technologies
Sari Merilampi, Krzysztof Iniewski, and Andrew Sirkka

PUBLISHED TITLES:

Ionizing Radiation Effects in Electronics: From Memories to Imagers
Marta Bagatin and Simone Gerardin

IoT and Low-Power Wireless: Circuits, Architectures, and Techniques
Christopher Siu

Labs on Chip: Principles, Design, and Technology
Eugenio Iannone

Laser-Based Optical Detection of Explosives
Paul M. Pellegrino, Ellen L. Holthoff, and Mikella E. Farrell

Low Power Emerging Wireless Technologies
Reza Mahmoudi and Krzysztof Iniewski

Low Power Semiconductor Devices and Processes for Emerging Applications in Communications, Computing, and Sensing
Sumeet Walia

Magnetic Sensors and Devices: Technologies and Applications
Kirill Poletkin and Laurent A. Francis

Medical Imaging: Technology and Applications
Troy Farncombe and Krzysztof Iniewski

Metallic Spintronic Devices
Xiaobin Wang

MEMS: Fundamental Technology and Applications
Vikas Choudhary and Krzysztof Iniewski

Micro- and Nanoelectronics: Emerging Device Challenges and Solutions
Tomasz Brozek

Microfluidics and Nanotechnology: Biosensing to the Single Molecule Limit
Eric Lagally

MIMO Power Line Communications: Narrow and Broadband Standards, EMC, and Advanced Processing
Lars Torsten Berger, Andreas Schwager, Pascal Pagani, and Daniel Schneider

Mixed-Signal Circuits
Thomas Noulis

Mobile Point-of-Care Monitors and Diagnostic Device Design
Walter Karlen

Multisensor Attitude Estimation: Fundamental Concepts and Applications
Hassen Fourati and Djamel Eddine Chouaib Belkhiat

Multisensor Data Fusion: From Algorithm and Architecture Design to Applications
Hassen Fourati

PUBLISHED TITLES:

MRI: Physics, Image Reconstruction, and Analysis
Angshul Majumdar and Rabab Ward

Nano-Semiconductors: Devices and Technology
Krzysztof Iniewski

Nanoelectronic Device Applications Handbook
James E. Morris and Krzysztof Iniewski

Nanomaterials: A Guide to Fabrication and Applications
Sivashankar Krishnamoorthy

Nanopatterning and Nanoscale Devices for Biological Applications
Šeila Selimović

Nanoplasmonics: Advanced Device Applications
James W. M. Chon and Krzysztof Iniewski

Nanoscale Semiconductor Memories: Technology and Applications
Santosh K. Kurinec and Krzysztof Iniewski

Noise Coupling in System-on-Chip
Thomas Noulis

Novel Advances in Microsystems Technologies and Their Applications
Laurent A. Francis and Krzysztof Iniewski

Optical, Acoustic, Magnetic, and Mechanical Sensor Technologies
Krzysztof Iniewski

Optical Fiber Sensors: Advanced Techniques and Applications
Ginu Rajan

Optical Imaging Devices: New Technologies and Applications
Ajit Khosla and Dongsoo Kim

Organic Solar Cells: Materials, Devices, Interfaces, and Modeling
Qiquan Qiao

Physical Design for 3D Integrated Circuits
Aida Todri-Sanial and Chuan Seng Tan

Power Management Integrated Circuits and Technologies
Mona M. Hella and Patrick Mercier

Radiation Detectors for Medical Imaging
Jan S. Iwanczyk

Radiation Effects in Semiconductors
Krzysztof Iniewski

Reconfigurable Logic: Architecture, Tools, and Applications
Pierre-Emmanuel Gaillardon

Semiconductor Devices in Harsh Conditions
Kirsten Weide-Zaage and Malgorzata Chrzanowska-Jeske

PUBLISHED TITLES:

Semiconductor Radiation Detection Systems
Krzysztof Iniewski

Semiconductor Radiation Detectors, Technology, and Applications
Salim Reza

Semiconductors: Integrated Circuit Design for Manufacturability
Artur Balasinski

Sensors for Diagnostics and Monitoring
Kevin Yallup and Laura Basiricò

Smart Grids: Clouds, Communications, Open Source, and Automation
David Bakken

Smart Sensors for Industrial Applications
Krzysztof Iniewski

Soft Errors: From Particles to Circuits
Jean-Luc Autran and Daniela Munteanu

Solid-State Radiation Detectors: Technology and Applications
Salah Awadalla

Structural Health Monitoring of Composite Structures Using Fiber Optic Methods
Ginu Rajan and Gangadhara Prusty

Technologies for Smart Sensors and Sensor Fusion
Kevin Yallup and Krzysztof Iniewski

Telecommunication Networks
Eugenio Iannone

Testing for Small-Delay Defects in Nanoscale CMOS Integrated Circuits
Sandeep K. Goel and Krishnendu Chakrabarty

Tunable RF Components and Circuits: Applications in Mobile Handsets
Jeffrey L. Hilbert

VLSI: Circuits for Emerging Applications
Tomasz Wojcicki

Wireless Medical Systems and Algorithms: Design and Applications
Pietro Salvo and Miguel Hernandez-Silveira

Wireless Technologies: Circuits, Systems, and Devices
Krzysztof Iniewski

Wireless Transceiver Circuits: System Perspectives and Design Aspects
Woogeun Rhee

FORTHCOMING TITLES:

Compressed Sensing for Engineers
Angshul Majumdar

Energy Efficient Computing: Devices, Circuits, and Systems
Santosh K. Kurinec and Sumeet Walia

Low Power Circuits for Emerging Applications in Communications, Computing, and Sensing
Krzysztof Iniewski and Fei Yuan

Radio Frequency Integrated Circuit Design
Sebastian Magierowski

Spectral Computed Tomography: Technology and Applications
Katsuyuki Taguchi, Ira Blevis, and Krzysztof Iniewski

X-Ray Diffraction Imaging: Technology and Applications
Joel Greenberg and Krzysztof Iniewski

For more information about this series, please visit:
https://www.crcpress.com/Devices-Circuits-and-Systems/book-series/CRCDEVCIRSYS

Compressed Sensing for Engineers

Angshul Majumdar

CRC Press is an imprint of the
Taylor & Francis Group, an **informa** business

CRC Press
Taylor & Francis Group
6000 Broken Sound Parkway NW, Suite 300
Boca Raton, FL 33487-2742

First issued in paperback 2022

ISBN-13: 978-0-815-36556-3 (hbk)
ISBN-13: 978-1-03-233871-2 (pbk)
DOI: 10.1201/9781351261364

Publisher's Note

The publisher has gone to great lengths to ensure the quality of this reprint but points out that some imperfections in the original copies may be apparent.

Library of Congress Cataloging-in-Publication Data

Names: Majumdar, Angshul, author.
Title: Compressed sensing for engineers / Angshul Majumdar.
Description: First edition. | Boca Raton, FL : CRC Press/Taylor & Francis, [2019] | Series: Devices, circuits, and systems
Identifiers: LCCN 2018037706| ISBN 9780815365563 (hardback : alk. paper) | ISBN 9781351261364 (ebook)
Subjects: LCSH: Compressed sensing (Telecommunication) | Image processing--Digital techniques. | Image compression. | Signal processing--Mathematics.
Classification: LCC TA1638 .M35 2019 | DDC 621.36/78--dc23
LC record available at https://lccn.loc.gov/2018037706

Visit the Taylor & Francis Web site at
http://www.taylorandfrancis.com

and the CRC Press Web site at
http://www.crcpress.com

Contents

Foreword

This is an appropriately timed book on the topic of compressed sensing: A subject of interest for slightly more than a decade now and one that has since demanded a treatise accessible to engineers. Only a few well-known books exist on the topic of compressed sensing and sparse recovery, and an understanding of those volumes requires comprehensive knowledge of advanced mathematics. This publication focuses uniquely on the basic mathematics engineers study during their undergraduate degree, explaining concepts intuitively and bypassing the esoteric mathematics of compressed sensing.

The book concentrates on the algorithms and applications of compressed sensing. The discussion begins with greedy-type algorithms for sparse recovery, followed by optimization-based techniques. Structured compressed sensing algorithms are discussed later on. Although not exactly classical compressed sensing, low-rank matrix recovery techniques are also addressed. This first portion of the book on algorithms concludes with a discussion of the state-of-the-art adaptive sparse representation learning techniques.

Compressed sensing has found application across various fields, ranging from scientific imaging, to biomedical signal processing, to analog-to-information hardware chips. It is hard to cover all these fields in a single volume; this book does a wonderful job of discussing the most prominent areas to which compressed sensing has been applied. These applications include computational imaging, medical imaging, biomedical signal processing, machine learning, and denoising.

One of the unique aspects of this book is the sheer volume of MATLAB® algorithms associated with it. Codes for almost all the algorithms discussed in this book have been made available and will serve as a valuable resource for practicing engineers.

This book is ideal for senior undergraduate and graduate students working in the areas of signal processing or machine learning. Course instructors may also apply it as source material. Today, compressed sensing, as a subject, is only taught in top universities, where both students and the instructors have the mathematical maturity to understand the topic from available resources. I am convinced that the publication of this book will change this engineering landscape, introducing greater understanding of compressed sensing to engineering masses.

Dr. Rabab K. Ward
Fellow, IEEE
Past President, IEEE Signal Processing Society
Professor Emeritus, Electrical and Computer Engineering
University of British Columbia

Preface

Compressed sensing deals with the topic of solving an under-determined system of linear equations when the solution is known to be sparse. Although an apparently esoteric topic, it has touched many practical facets of science of engineering in slightly over a decade of its being.

The idea of using l_1-norm or l_0-norm as a sparsity penalty has been known since the 1970s; however, the thorough theoretical understanding was possible only with the breakthroughs of Donoho, Candes, and Tao in the mid and late 2000s. It was during these years that the term compressed sensing/compressive sampling was coined. The rest, as they say, is history.

Compressed sensing is a fairly theoretical area. A large volume of articles on this topic revolved around proofs, tighter bounds, stability in the presence of perturbations, etc. An equally large volume of literature is on the topic of compressed sensing algorithms. These include various greedy techniques, orthogonal matching pursuit and its variants, as well as optimization-based methods such as spectral projected gradient algorithm and Nesterov's algorithm, to name a few. There are many papers on relations and extensions of compressed sensing to other areas of signal processing, such as finite rate of innovation, union of subspace, and model selection. To understand these papers, one needs at least a graduate-level degree in mathematics.

The success of compressed sensing did not hinge on the solid theoretical framework; it was because of its widespread applicability. Compressed sensing unleashed a new era of accelerating magnetic resonance scans, especially at a juncture when it was believed that "whatever could be done has been done" on this front. It ushered in safer computed tomographic (CT) scans with lower doses of X-radiation. It was used in analyzing DNA microarrays. It was used by geosciences to fasten wavefield extrapolation—compressed sensing was able to reduce the time by an order of magnitude (from weeks to days). It was used in radar imaging to improve resolution, without increasing any other costs. There are many other areas of applied sciences that benefit from compressed sensing, but it is not possible to mention all of them. However, we believe one gets the idea on its range of applicability.

Despite its success, unfortunately, there is no unified resource for understanding this topic at the basic level. There are two books on this topic that are fairly mathematical and well beyond the reach of a graduate or practicing engineer. The third book is an anthology, which loosely combines a motley of articles from top researchers in the field. None of them caters to regular engineers, who might not have a degree in advanced mathematics.

The author of this book felt this lacuna when he started teaching this course to graduate and senior undergraduate students at his university in 2014. As an instructor, the author realized that it is almost impossible to teach fundamental concepts of compressed sensing such as the restricted isometric property or the null space property to a student who is finishing/has just finished his/her first degree in engineering. However, making the students understand the topics intuitively was fairly simple. This led the author to write this book—a concise treatise on the topic of compressed sensing meant for engineers worldwide.

We assume that the reader of this book will have a basic understanding of mathematics. He/she would have taken the first courses in linear algebra and some numerical techniques. This is not too much of an ask, given the fact that engineers all around the world are mandated to take these courses in their undergraduate education.

In this book, we have thoughtfully left aside the theory of compressed sensing. This is largely because there are several books on this topic. The concentration of this book is on algorithms and applications—topics that have not been covered adequately by others.

The first portion of the book is on the algorithms. We start with algorithms for classical compressed sensing and then branch off to its extensions such as group sparsity and row sparsity. Although not exactly compressed sensing, we discuss the topic of low-rank matrix recovery, since the two are related. Finally, we cover the topic of dictionary learning.

The second portion of the book is about applications. Admittedly, we fail to cover every application of compressed sensing. The book tries to cover some of the major areas such as computational imaging, medical imaging, biomedical signal processing, and machine learning.

This is not a textbook for compressed sensing. But we believe that it can serve as a ready material for developing the first course on the topic. A large number of MATLAB® examples have been provided to help the reader get a head start on compressed sensing.

Angshul Majumdar

MATLAB® is a registered trademark of The MathWorks, Inc. For product information, please contact:

The MathWorks, Inc.
3 Apple Hill Drive
Natick, MA 01760-2098 USA
Tel: 508 647 7000
Fax: 508-647-7001
E-mail: info@mathworks.com
Web: www.mathworks.com

Acknowledgments

I thank my PhD supervisor, Dr. Rabab Ward, for making me what I am today. She has received numerous accolades in her illustrious career, but these awards do not capture the life-changing influence she has on her students. Unlike most good supervisors, she does not stop at defining the research problem and technically helping them if need arises—she encouraged us to think out of the box and to explore areas beyond the convention. I can only try to pass on her virtues to my students.

I am indebted to Dr. Felix Herrmann of UBC for introducing me to the topic of compressed sensing. It was my first year as a graduate (master's) student in fall 2007. Compressed sensing is a very mathematical topic, but Felix introduced the subject in a fashion that I could easily grasp with an engineer's background in mathematics. He interpreted the mathematical results in a very intuitive and interesting fashion. Throughout the book, I have tried to follow Felix's philosophy of explaining the mathematical nuances in an easy-to-understand way.

Perhaps the biggest source of inspiration for writing this book came from my students of compressed sensing course, especially the first MTech batch who I taught in 2014. Over the years, I received constructive comments from the students which helped me shape the course. The book is largely based on my course that evolved over the last four years.

I am thankful to Dr. Dipti Prasad Mukherjee, professor at Indian Statistical Institute (ISI), for hosting me as a visiting faculty every summer for the last three years. This allowed me to get away from my day-to-day research and concentrate solely on writing this book. Without ISI's facilities, this book would have never been completed.

Finally, I thank Dr. Pankaj Jalote, the founding director of Indraprastha Institute of Information Technology, Delhi, for allowing me the freedom to teach such an advanced course. This was one of the first courses in compressed sensing in the country. Without this freedom, I would not have been able to design the course and thereby the book.

Acknowledgments

Author

Angshul Majumdar, PhD, earned a BE in electronics and communication engineering from the Bengal Engineering and Science University, Shibpur, India, in 2005, and an MASc and PhD in electrical and computer engineering from the University of British Columbia, Vancouver, Canada, in 2009 and 2012, respectively. Since 2012, he is an assistant professor at Indraprastha Institute of Information Technology, Delhi, India. Prior to his current position, he worked with PricewaterhouseCoopers India as a senior consultant in technology advisory. His current research interests include signal processing and machine learning. Dr. Majumdar has co-authored more than 150 articles in top-tier journals and for conferences. He has written a book titled *Compressed Sensing Based Magnetic Resonance Imaging Reconstruction,* published by Cambridge University Press (2015). He has co-edited two books: *MRI: Physics, Reconstruction and Analysis* (2015) and *Deep Learning in Biometrics* (2018), both published by CRC Press.

1

Introduction

There is a difference between data and information. We hear about big data. In most cases, the humongous amount of data contains only concise information. For example, when we read a news report, we get the gist from the title of the article. The actual text has just the details and does not add much to the information content. Loosely speaking, the text is the data and the title is its information content.

The fact that data is compressible has been known for ages; that is, the field of statistics—the art and science of summarizing and modeling data—was developed. The fact that we can get the essence of millions of samples from only a few moments or represent it as a distribution with very few parameters points to the compressibility of data.

Broadly speaking, compressed sensing deals with this duality—abundance of data and its relatively sparse information content. Truly speaking, compressed sensing is concerned with an important sub-class of such problems—where the sparse information content has a linear relationship with data. There are many such problems arising in real life. We will discuss a few of them here so as to motivate you to read through the rest of the book.

Machine Learning

Consider a health analyst trying to figure out what are the causes of infant mortality in a developing country. Usually, when an expecting mother comes to a hospital, she needs to fill up a form. The form asks many questions: for example, the date of birth of the mother and father (thereby their ages), their education level, their income level, type of food (vegetarian or otherwise), number of times the mother has visited the doctor, and the occupation of the mother (housewife or otherwise). Once the mother delivers the baby, the hospital keeps a record of the baby's condition. Therefore, there is a one-to-one correspondence between the filled-up form and the condition of the baby. From this data, the analyst tries to find out what are the factors that lead to the outcome (infant mortality).

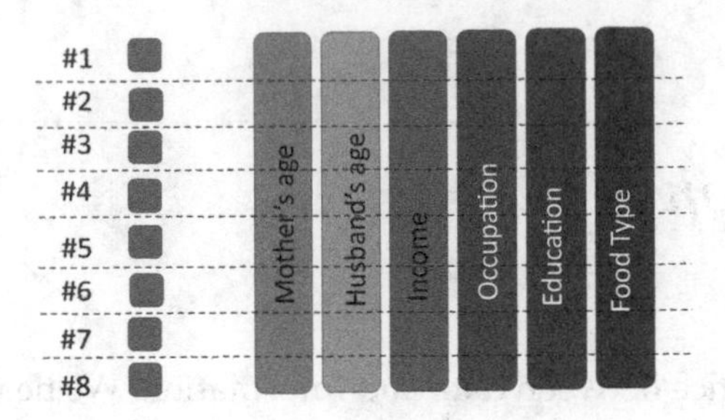

FIGURE 1.1
Health record and outcome.

For simplicity, the analyst can assume that the relationship between the factors (age, health, income, etc.) and the outcome (mortality) is linear, so that it can be expressed as shown in the following (Figure 1.1).

Formally, this can be expressed as follows:

$$b = Hx + n \tag{1.1}$$

where b is the outcome, H is the health record (factors along columns and patients along rows), and x is the (unknown) variable that tells us the relative importance of the different factors. The model allows for some inconsistencies in the form of noise n.

In the simplest situation, we will solve it by assuming that the noise is normally distributed; we will minimize the l_2-norm.

$$x = \min_x \|b - Hx\|_2^2 \tag{1.2}$$

However, the l_2-norm will not give us the desired solution. It will yield an x that is dense, that is, will have non-zero values in all positions. This would mean that ALL the factors are somewhat responsible for the outcome. If the analyst says so, the situation would not be very practical. It is not possible to control all aspects of the mother (and her husband's) life. Typically, we can control only a few factors, not all. But the solution (1.2) does not yield the desired solution.

Such types of problems fall under the category of regression. The simplest form of regression, the ridge regression/Tikhonov regularization (1.3), does not solve the problem either.

$$x = \min_x \|b - Hx\|_2^2 + \lambda \|x\|_2^2 \tag{1.3}$$

This too yields a dense solution, which in our example is ineffective. Such problems have been studied for long statistics. Initial studies in this area proposed greedy solutions, where each of the factors was selected following some heuristic criterion. In statistics, such techniques were called sequential forward selection; in signal processing, they were called matching pursuit.

The most comprehensive solution to this problem is from Tibshirani; he introduced the Least Angle Selection and Shrinkage Operator (LASSO). He proposed solving,

$$x = \min_x \|b - Hx\|_2^2 + \lambda \|x\|_1 \tag{1.4}$$

The l_1-norm penalty promotes a sparse solution (we will learn the reason later). It means that x will have non-zero values only corresponding to certain factors, and the rest will be zeroes. This would translate to important (non-zero) and unimportant (zero) factors. Once we decide on a few important factors, we can concentrate on controlling them and improving child mortality.

However, there is a problem with the LASSO solution. It only selects the most important factor from the set of related factors. For example, in child's health, it is medically well known that the parents' age is an important aspect. Therefore, we consider the age of both parents. If LASSO is given a free run (without any medical expertise in deciding the factors), instead of pointing out to both parent's ages, it will select only the most relevant one—say the mother's age.

If we are to decide the factors solely based on data analysis (and no medical knowledge), trying to control the outcomes based on the factors' output by LASSO would not always yield the correct/expected results. LASSO will leave aside related important factors. To combat this issue, the elastic-net regularization was proposed. It is of the form,

$$x = \min_x \|b - Hx\|_2^2 + \lambda_1 \|x\|_1 + \lambda_2 \|x\|_2 \tag{1.5}$$

In the elastic-net regularization, the l_1-norm and l_2-norm have opposing effect. The l_1-norm enforces a sparse solution, whereas the l_2-norm promotes density. Owing to these opposing effects, it is called "elastic" net. The net effect is that it selects only a few factors, but the grouping effect of l_2-norm selects all the correlated factors as well.

Sometimes, it is known beforehand that some of the factors will always happen together, for example, mother's and father's age. To a large extent, education and income are also related. Such information cannot be exploited in the LASSO or elastic-net framework. This requires the group-LASSO formulation, where the factors are grouped and happen simultaneously. The formulation is,

$$x = \min_x \|b - Hx\|_2^2 + \lambda \|x\|_{2,1} \tag{1.6}$$

The $\|x\|_{2,1}$ is defined as the sum of the l_2-norm over the groups. The l_2-norm within the group promotes a dense solution within the group; that is, if the group is selected, all the factors in the group will have non-zero values; but the sum over the l_2-norms acts as an outer l_1-norm and enforces sparsity in the selection of groups. Such problems belong to the category of structured sparse problems.

Signal Processing

Similar problems arise in signal processing. For example, in functional magnetic resonance imaging (fMRI), we image section of the brain, usually the blood-oxygen-level-dependent (BOLD) signal. The part of the brain that shows activity has more blood oxygen, so that portion lights up in fMRI. Typically, only a small portion is activated, and hence, the signal is sparse in spatial domain. MRI is acquired in the Fourier frequency space; for historical reasons, this is called the K-space.

MRI is a slow imaging modality; it has great spatial resolution but very poor temporal resolution. Therefore, only 6–10 fMRI frames can be captured per second. Owing to poor temporal resolution, transient (short-term) effects cannot be studied by conventional fMRI. Therefore, a significant effort is ongoing to reduce the scan time. This can be reduced only if we capture the Fourier frequency space partially. Mathematically, this is expressed as,

$$y = RFx + n \tag{1.7}$$

Here, x is the underlying image; F is the Fourier transform; R is the restriction/masking operator denoting under-sampling; and y is the acquired Fourier frequency samples. This is an under-determined problem. This is because Fourier transform is orthogonal, but since it is only partially acquired, the size of y is much smaller than the size of x, making the inverse problem (1.7) under-determined.

However, in this particular scenario, we know that x is sparse; this is because only a portion of the brain is active, which leads to non-zero-valued pixels, while other areas with no/negligible activity are zeroes or close to zero. This makes (1.7) an ideal candidate for compressed sensing reconstruction. The image is recovered as,

$$\min_x \|y - RFx\|_2^2 + \lambda \|x\|_1 \tag{1.8}$$

The acquisition formulation (1.7) is true for MRI scanners having a single receiver channel. In most cases, multiple receivers are used for parallel acquisition. In that case, the acquisition is expressed as,

$$y_i = RFx_i + n \tag{1.9}$$

where 'i' denotes the channel number. Here, $x_i = S_i x$ is the sensitivity encoded image, as "seen" by the ith scanner.

This is a multiple measurement vector problem. It can succinctly be represented as,

$$Y = RFX + N \tag{1.10}$$

Here, Y and X are created by stacking the y_is and x_is as columns, respectively. In this formulation, X has a special structure—it is row-sparse. This is because all the receivers will essentially see the same underlying image; the positions of the non-zero values are going to remain the same. This would mean, only those rows in x corresponding to the active positions of the brain will have non-zero values. Therefore, the Xs need to be recovered by joint/row-sparse recovery.

$$\|Y - RFX\|_F^2 + \lambda \|X\|_{2,1} \tag{1.11}$$

Here, the $l_{2,1}$-norm has a different connotation from (1.6). However, there should not be any confusion between the two, since it is defined over a matrix in (1.11), whereas it was defined over a vector in (1.6). In this case, the $l_{2,1}$-norm is defined as the sum of the l_2-norms of the rows. The idea remains the same as before. The outer sum of portion behaves like an l_1-norm and selects only a few rows, whereas the inner 12-norm promotes dense solution within the selected row.

Low-Rank Recovery

So far, we kept talking about sparse recovery and associated problems. Truly speaking, this topic forms the crux of compressed sensing. However, the problem of low-rank recovery has very similar traits to that of compressed sensing. In low-rank recovery, the solution is supposed to be low-rank; this is not sparse in itself; however, it has sparse singular values. A lot of theoretical studies on these two topics bear semblance with each other; they were developed by the same set of people. Therefore, in this book, we will broaden the definition of "compressed sensing" and include low-rank matrix recovery as well. We will see why this problem is important from a couple of problems.

Signal Processing

Consider the problem of dynamic medical imaging; here, frames are captured over time. The generic acquisition problem can be expressed as,

$$y_t = Ax_t + n \tag{1.12}$$

where A is the Fourier transform for MRI or the radon transform for X-ray computed tomography (CT).

The problem is to recover the frames. This can be done one at a time; however, such a piecemeal reconstruction would not optimally capture the spatio-temporal correlation in the sequence x_t. In most cases, the entire sequence is reconstructed simultaneously. The problem (1.12) is posed as,

$$Y = AX + N \tag{1.13}$$

The symbols have their usual meaning. The goal is to recover the sequence of images X. In any dynamic video sequence, there is significant spatial and temporal correlation between the frames. Therefore, the matrix X will be low-rank, since its columns will not be linearly dependent. This allows (1.13) to be solved by using low-rank matrix recovery techniques. The traditional way to solve the problem is to guess the rank of the matrix X and express it as a product of two matrices, U and V. Therefore, (1.13) can be expressed as,

$$Y = AUV + N \tag{1.14}$$

This is solved by alternating updating U and V. This technique is called matrix factorization. This is a non-convex problem. A better way to solve the low-rank recovery problem is via nuclear-norm minimization.

$$\min_X \|Y - AX\|_F^2 + \lambda\|X\|_* \tag{1.15}$$

The nuclear norm is defined as the sum of nuclear values and is the tightest convex surrogate for matrix rank. This is a convex problem and enjoys better theoretical guarantees than matrix factorization.

Machine Learning

Most of us have used Amazon, and almost all of us have seen YouTube. When we login to these portals and see something, we get some recommendations regarding related products/videos. The technique for recommending these is called "recommender system." Today, the de facto standard for recommender system is collaborative filtering.

Recommendations are based on user's ratings on items. These can be explicit ratings; many a time, we are asked to rate movies and videos; after a

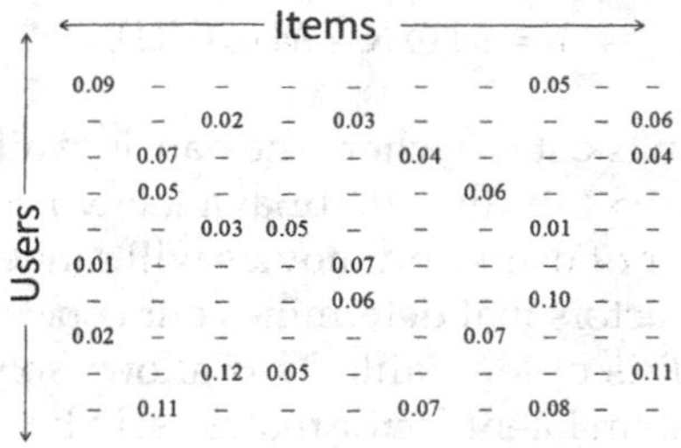

FIGURE 1.2
Ratings matrix.

purchase on Amazon, we are asked to rate the product once it is delivered. Ratings can also be implicit, based on user's browsing history or purchase pattern. In general, the ratings can be expressed as a matrix—one side of it represents the users (say rows) and the other side the items (say columns) (Figure 1.2).

Consider a concrete example of movie ratings. Our choice of movies is determined by a few factors—genre, actors, director, etc. However, we do not know for certain how many such factors are there and what they are. Therefore, latent factor model assumes an abstract set of factors to shape our choice. Each person is characterized by a vector of such latent factors; basically, this tells the user's propensity toward these factors. Movies are also characterized by the same set of factors—they either possess these or not. A person will like a movie if the factors he/she desires are present in the movie. A simple way to express it is via the inner product of the user and item latent factors, that is,

$$r_{i,j} = v_i u_j^T \tag{1.16}$$

Here, $r_{i,j}$ is the ith user's rating on the jth movie; v_i is the latent factor vector of user i; and u_j is the latent factor vector of movie j. When written in terms of all users and items, this is expressed as,

$$R = VU \tag{1.17}$$

where R is the complete ratings matrix, V is the latent factor matrix for users, and U is the latent factor matrix for movies.

In collaborative filtering, R is never fully available. If it were, the problem would not exist. If we knew every user's rating on every movie, we would only give or recommend the highly rated movies. The matrix R is always incomplete, because the user never rates all movies. This is partly because one does not want to rate millions of movies and partly because he/she has not seen all movies. Therefore, in practical scenarios, the rating matrix is only partially available. This is expressed as,

$$Y = M \odot R = M \odot (VU) \tag{1.18}$$

Here, M is the binary mask; it is 1 where the data is available and 0 otherwise. Typical problems have 95% of the data unavailable. Y is the acquired rating.

Typically, the number of users and movies will range in hundreds of thousands. The number of factors that determines our choice is very few (40 or so); therefore, the matrix R is of low rank. This allows solving for R directly or through the user and item latent factor matrices; in both cases, it turns out to be a low-rank matrix recovery problem.

2

Greedy Algorithms

Let us recapitulate the problem; we have to solve the following under-determined linear inverse problem,

$$y_{m\times1} = A_{m\times n}x_{n\times1}, m < n \tag{2.1}$$

Since this is an under-determined inverse problem, in general, it has infinite number of solutions. Our interest is in a sparse solution. By sparse, we mean that the n dimensional vector (x) has only s non-zeroes and rest n-s zeroes; we will call such vectors to be s-sparse. Is it possible to solve for an s-sparse vector by solving the under-determined inverse problem (2.1)?

Oracle Solution

Let us try to understand the problem intuitively. In the most favorable scenario (say an oracle), we would know the s positions corresponding to the s non-zeroes in x. Let Ω be the set of non-zero indices. In such a case, solving (2.1) is easy. We pick only the columns indexed Ω; let it be denoted as A_Ω. Assuming that the number of equations (length of y) is larger than s, we have to solve only the following over-determined problem,

$$y = A_\Omega x_\Omega \tag{2.2}$$

Here, x_Ω denotes that we are solving only for those positions in x that are known to have non-zero values (given by Ω). The problem (2.2) is over-determined and therefore has a left inverse. Thus, the non-zero positions in x can be obtained by,

$$x_\Omega = \left(A_\Omega^T A_\Omega\right)^{-1} A_\Omega^T y \tag{2.3}$$

Once the values at the non-zero positions are obtained, the rest of the positions in x are filled with zeros.

(Im)Practical Solution

Unfortunately, the positions of the non-zero elements in x will not be known. In such a situation, a brute force technique can be applied to solve (2.1). We only know that the solution is s-sparse. Since we do not know Ω, it is possible to choose s elements from n in ${}^nC_s$ ways; thus, there will be ${}^nC_s$ such Ωs. What we do is to solve (2.2) for every such Ω.

Mathematically, this problem is posed as,

$$\min_x \|x\|_0 \leq s \text{ such that } y = Ax \tag{2.4}$$

Strictly speaking, $\|x\|_0$ is not a norm; it only counts the number of non-zeroes in the vector.

One can see that this is a combinatorial problem. There is no polynomial time algorithm to solve (2.1) in such a situation. Hence, it is a non-deterministic polynomial (NP) hard problem. There is no technique that is smarter than the brute force method just discussed, and all have a combinatorial complexity. Then, how do we solve it?

Intuitive Solution

We consider a very simple problem: where the solution has only one non-zero values, that is, the solution is 1-sparse. As earlier, we need to solve the inverse problem:

$$y = Ax$$

In the most conducive scenario, A is orthogonal, so that its inverse is the transpose. Then, we could easily recover x by hitting both sides of the equation by A^T, that is,

$$A^T y = A^T Ax = x, \because A^T A = I \tag{2.5}$$

In the most common scenario, the system of equations will not be orthogonal. But what if they are "nearly"/"approximately" orthogonal? To take a concrete example, consider a restricted Fourier operator, that is, it is not the full Fourier matrix and has some of its rows (randomly selected) missing. In such a case, if we hit the inverse problem by the transpose of A (in our case restricted Fourier operator), we see something interesting, as in Figure 2.1.

The solution x had only a non-zero value at position 2; this is shown in by 'o' in Figure 2.1. After hitting the inverse problem by the transpose, that is,

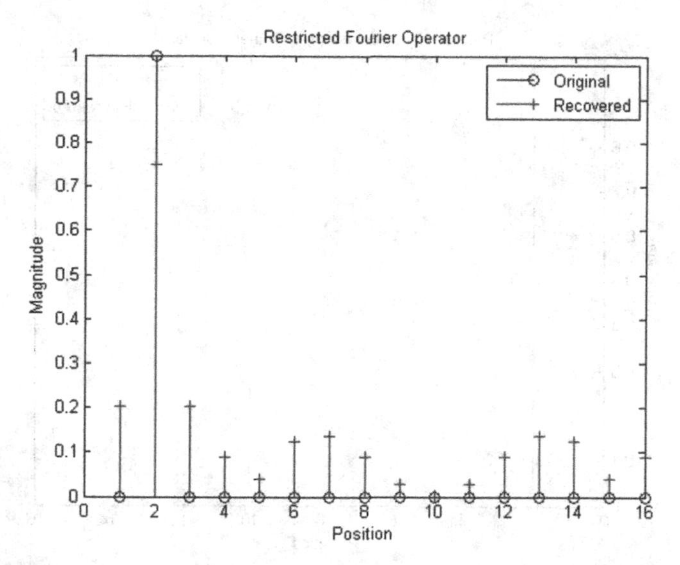

FIGURE 2.1
Approximate recovery for restricted Fourier operator.

$x_R = A^T y = A^T Ax$, we plot the magnitude of x_R in red. Note that the position of the highest-valued coefficient in x_R is exactly the same as the position of the non-zero coefficient in the exact solution x.

If the full Fourier matrix were available, it would be an orthogonal system and we would get the exact recovery. But we have this "approximate" scenario, where some of the rows are missing at random. Had the rows been missing at periodic intervals, we would not get this; rather, we would have folding/aliasing effects, as can be seen in Figure 2.2. Instead of having one high-valued coefficient, Figure 2.2 has two. In such a case, we cannot figure out the index of the high-valued coefficient in the solution.

The two examples are shown to corroborate that there are some situations where we will be able to recover the correct index of the sparse coefficient and other situations where we would not be. There is a lot of theory behind the recovery criteria, but at this stage, we do not need to go into those. Rather, let us try to understand intuitively what is happening.

If the Fourier operator is undersampled, there will be aliasing—no escape to that. For periodic undersampling, we are aware of the well-known aliasing effects; this is structured aliasing. In the first scenario, when the Fourier operator has randomly missing rows, aliasing is happening, but now, the aliasing is unstructured and hence appears as noise (Figure 2.1)—there are non-zero values at all positions, but the values are small and hence do not meddle with the high value at the correct position.

The same phenomenon is observed for several other kinds of matrices. In Figure 2.3, we do the same exercise with a Gaussian independent and identically distributed (i.i.d) matrix. The vector obtained after hitting the

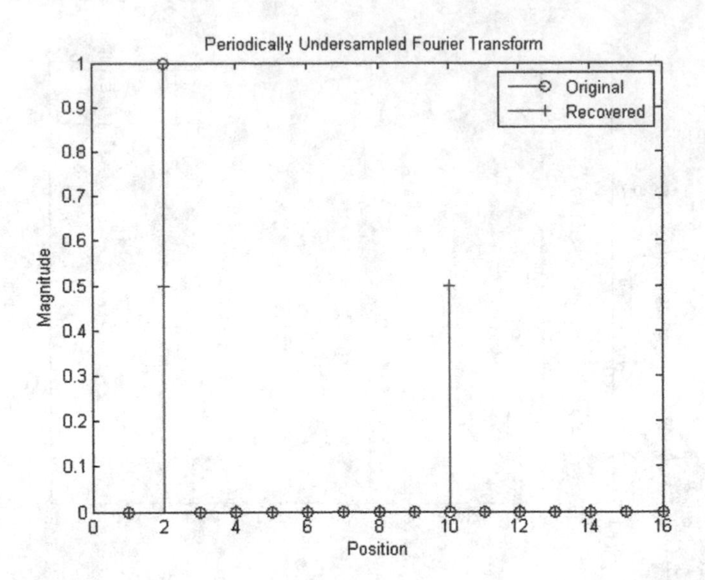

FIGURE 2.2
Aliasing with periodic undersampling.

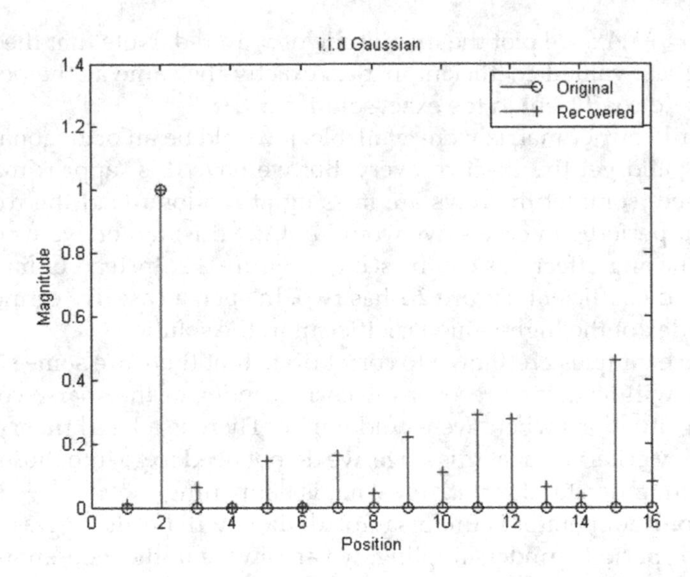

FIGURE 2.3
Approximate recovery for i.i.d Gaussian matrix.

system of equations with the transpose has a lot of non-zero coefficients, but the position of the highest-valued coefficient exactly coincides with the position of the non-zero value in the original solution.

These operators, where the transpose behaves approximately as an inverse, are good for us. They are said to satisfy the famous restricted isometry property (RIP). There are various other conditions under which solutions can be guaranteed, but we will not go into them.

So far, we have discussed how to find the position of the non-zero coefficient, but we have not yet found out the value at that position. Finding the value is simple and has already been discussed earlier in the section on "Oracle Solution." We just pick up the corresponding column of A and compute the left pseudoinverse to obtain the value at the required position.

Practical Solutions

In the previous section, we discussed a simple case where there is only one non-zero value in the solution, but this not of much practical importance. In general, we will be interested in solving problems where there are multiple non-zero values. How do we do that?

Matching Pursuit

The simplest algorithm is the matching pursuit (MP). The algorithm is as follows:

MATCHING PURSUIT

Initialize: $\Lambda = \{\}$ and $x = 0$.
Until convergence repeat:
Compute: $c = A^T\left(y - Ax_k\right)$
Detect support: $l^{(k)} = \arg\max_j |c_j|$
Append to existing support: $\Lambda = \Lambda \cup l^{(k)}$
Estimate solution: $x^{(k+1)} = x^{(k)} + c_{l(k)}$

In the first iteration, x is 0. Therefore, c is computed as the inner product of A and y. In the previous section, we discussed how for certain favorable matrices A^Ty preserve the position of the highest-valued coefficient. Thus, the first position is obtained by picking up the index with the highest magnitude in c. This is stored in $l^{(0)}$. Thus, $l^{(0)}$ is the first chosen index; it is added to the set of non-zero indices Λ.

MP is a computationally cheap algorithm. It does not find the values of the non-zero indices in the solution (x) by pseudoinverse; it simply takes the value of the highest coefficient (the same one used for finding the index) as the value of x at that position. Thus, after the first iteration, one of the positions in x is populated.

We do not want to get the same index again and again. Thus, once we have found an index, we try to eliminate it in the following iterations. That is why, after the first iteration, instead of computing $A^T y$, we compute $c = A^T(y - Ax)$. We compute the highest-valued index of c, add it to the support Λ, update x, and continue.

However, the simple computation of x based on the values of c does not orthogonalize the residual $y - Ax$; thus, there is always a possibility that the same index will be chosen again in some subsequent iteration. This problem is eliminated in orthogonal matching pursuit (OMP).

Orthogonal Matching Pursuit

To overcome the issue in MP, the OMP was proposed. OMP is more in tune with our earlier discussion, but there, we only learnt how to recover only 1-sparse solution. OMP generalizes it to the general s-sparse case.

OMP estimates the support (position of the non-zero values) iteratively. We show the case for a 2-sparse signal; generalizing it to the s-sparse case is trivial. Suppose we have a 2-sparse signal, as shown in Figure 2.4.

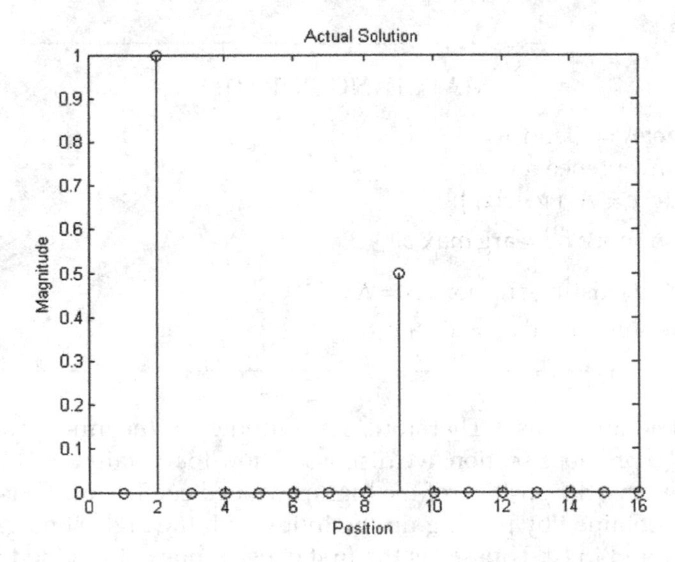

FIGURE 2.4
2-sparse solution.

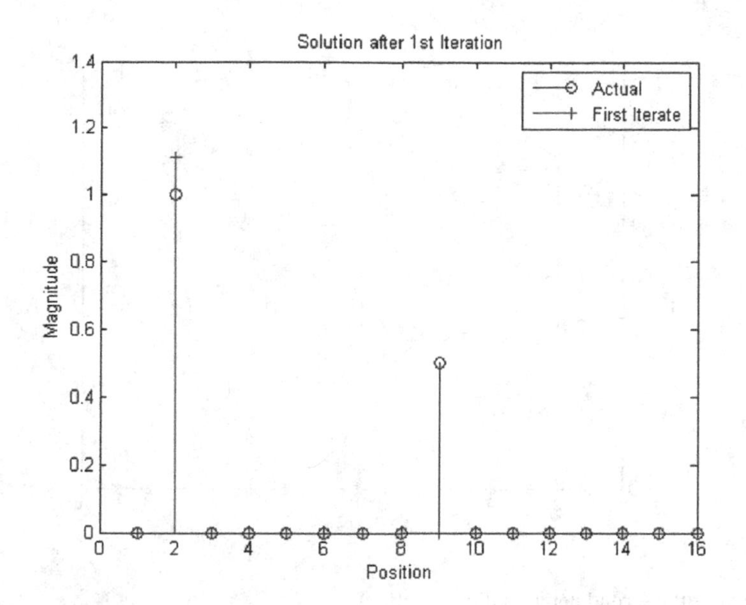

FIGURE 2.5
Solution after the first iteration.

We assume that A is an i.i.d Gaussian matrix. In the first iteration, we compute $c = A^T y$ and select the position of the highest value as the first index. The selected index is used to extract that column from A, and its pseudoinverse is computed to generate the first iterate. This is shown in Figure 2.5.

As can be seen, we have correctly estimated the support, but the value of x at the selected index is off. However, we have nothing to worry, as this will be corrected later.

In the second iteration, the residual is computed as $r = y - Ax^{(1)}$. Since the x is computed via pseudoinverse, the residual is orthogonalized; therefore, we do not choose the same position ever again. This fixes the problem arising in matching pursuit.

The residual is hit with A^T to obtain $c = A^T r$; the position of the highest value of c is added to the support set. Only those columns in A that are in the support set are extracted, and the pseudoinverse of the selected columns is used to generate the solution. This is shown in Figure 2.6. The correct solution is obtained.

In this case, since we knew that there only two non-zero elements, we stop after the second iteration. But, for larger number of non-zero coefficients, this procedure is repeated.

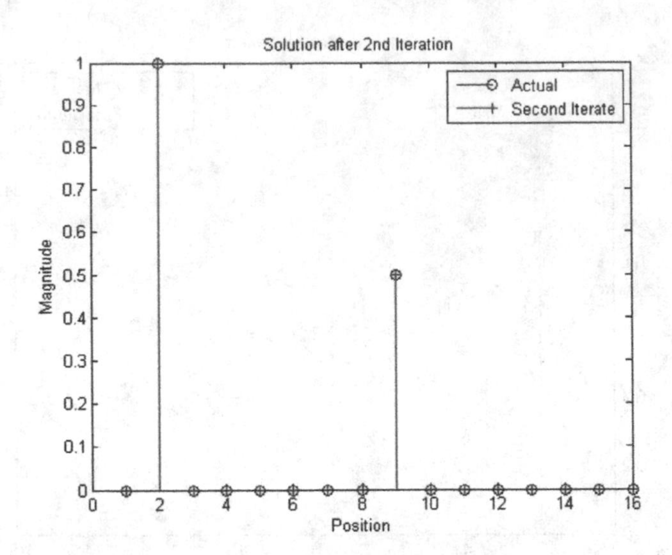

FIGURE 2.6
Solution after the second iteration (final solution).

In a concise fashion, OMP can be represented as follows:

> **ORTHOGONAL MATCHING PURSUIT**
>
> Initialize: $\Lambda = \{\}$ and $x = 0$.
> Until convergence repeat:
> Compute: $c = A^T\left(y - Ax_k\right)$
> Detect support: $l^{(k)} = \arg\max_j \left|c_j\right|$
> Append to existing support: $\Lambda = \Lambda \cup l^{(k)}$
> Estimate solution: $x^{(k+1)} = \min_x \left\|y - A_\Lambda x_\Lambda\right\|_2^2$

Here, x is estimated via least squares. This guarantees that $x^{(k+1)}$ is orthogonal to the residual $y - Ax^{(k+1)}$; thus, any index that has already been chosen will be selected in some further iteration. The price to be paid by OMP is a slightly more computation compared with the MP.

Stagewise Orthogonal Matching Pursuit

OMP detects one support in every iteration. This is slow for most large-scale practical problems. Consider a simple imaging problem: a 256 × 256 image consists of 65536 pixel values; assuming the image is sparse in wavelet

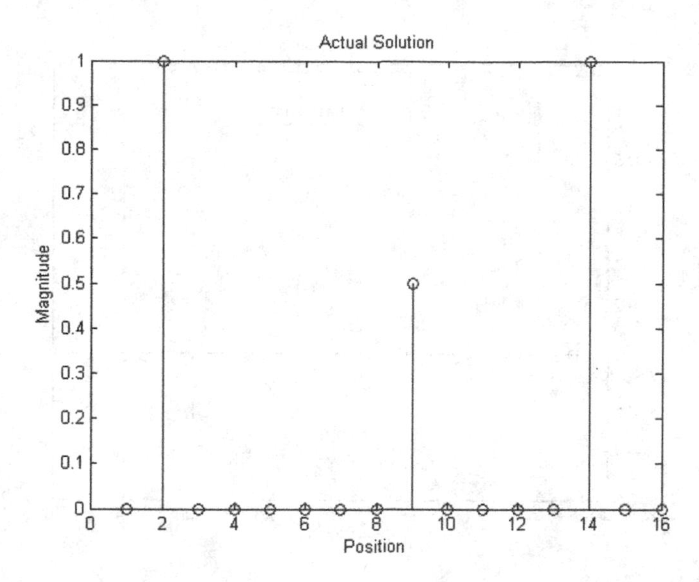

FIGURE 2.7
Actual solution—3-sparse.

domain, solving for even 10% of the non-zero wavelet coefficients (~6500) will take forever if OMP is used. There have been a lot of recent studies to speed up pursuit algorithms by selecting multiple supports in every iteration.

One of the simplest methods to do so is the stagewise OMP (StOMP). Instead of selecting one index at every time, StOMP selects all indices whose values are above a pre-determined threshold. The previous support is updated based on the new indices, and as before (in OMP), the signal values of the given support is computed via pseudoinverse.

Before going into the algorithm, let us understand what is going on. Take an example where there are three non-zero coefficients in the solution. This is shown in Figure 2.7. The non-zero values are at 2, 9, and 14.

Assuming that the system is i.i.d Gaussian, the vector we obtain after $A^T y$ is as shown in Figure 2.8 (superimposed on original solution). As can be seen, two of the highest-valued coefficients in $A^T y$ coincide with the actual support.

In OMP, we would have selected only the index corresponding to the highest magnitude. StOMP is greedier; it will select all indices over a pre-defined threshold. Thus, after thresholding, we get indices 2 and 14, and this is added to the hitherto empty support set. The values at the current support are obtained via pseudoinverse (as in OMP). This leads to the following solution after the first iteration. As we have seen in OMP, the values are a bit off at the estimated supports, but this will be rectified in subsequent iterations (Figure 2.9).

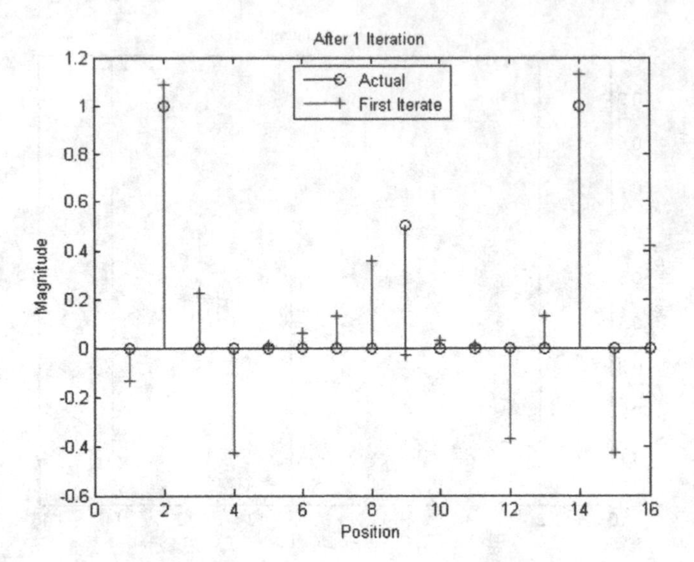

FIGURE 2.8
After first iteration.

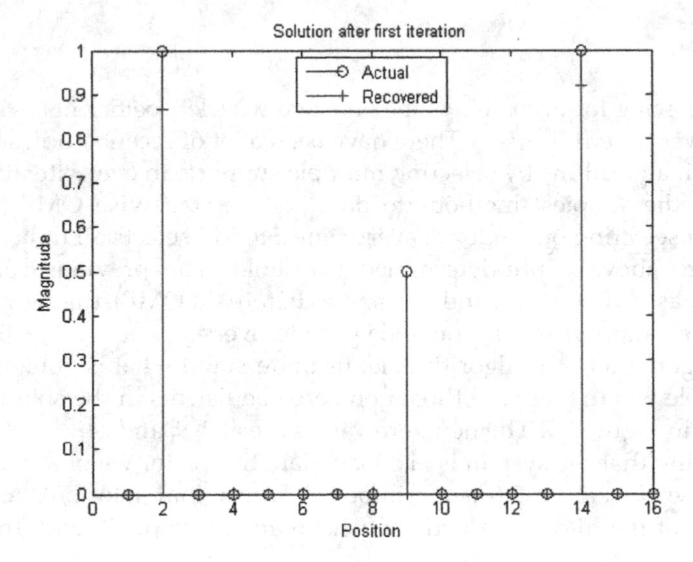

FIGURE 2.9
Solution after first iteration.

In the second iteration, the residual is computed as $r = y - Ax^{(0)}$. The plot of $A^T r$ versus the actual solution is shown in Figure 2.10. As can be seen, this time, the highest value is exactly at the remaining position – 9. Also notice that the values at 2 and 14 (the support detected in the first iteration) are exactly zero—this is owing to the orthogonalization. Any previously selected index will not be reselected via OMP type methods.

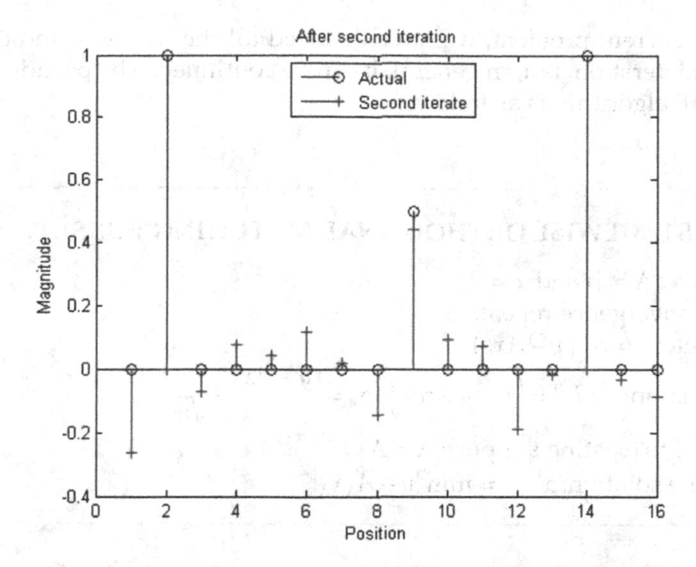

FIGURE 2.10
Solution after second iteration.

We follow the same procedure as before. By choosing a proper threshold, we find the support from the second iterate—this turns out to be 9. This index is added to the existing support (2 and 14); the values of x at the current support are estimated via pseudoinverse. This time, the values of x at all the non-zero positions come out correct (Figure 2.11).

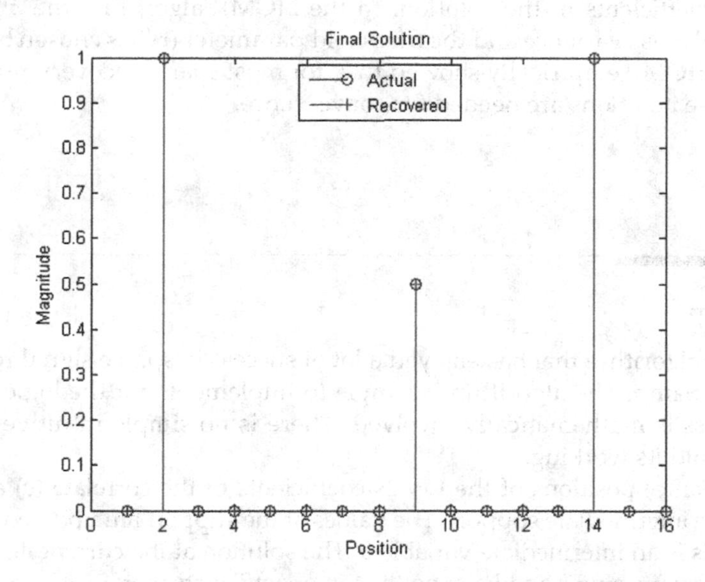

FIGURE 2.11
Final solution after second iteration.

For the current problem, we have detected all the non-zero indices after the second iteration. But, in general, it can be continued. The pseudocode for the StOMP algorithm is as follows:

STAGEWISE ORTHOGONAL MATCHING PURSUIT

Initialize: $\Lambda = \{\}$ and $x = 0$.
Until convergence repeat:
Compute: $c = A^T\left(y - Ax_k\right)$
Detect support: $l^{(k)} = \left\{j : \left|c_j\right| > \tau\sigma_k\right\}, \sigma_k = \left\|y - Ax_k\right\|_2 / \sqrt{n}$
Append to existing support: $\Lambda = \Lambda \cup l^{(k)}$
Estimate solution: $x^{(k+1)} = \min_x \left\|y - A_\Lambda x_\Lambda\right\|_2^2$

StOMP combines ideas from greedy algorithms such as OMP and the soft thresholding technique used for denoising signals. We have argued before that, since the system behaves to be approximately orthogonal, the positions of the high-valued coefficients are preserved and the positions of the zero-valued coefficients are corrupted by noise (from energy leakage). StOMP treats the support selection problem like denoising; by choosing a suitable threshold, it gets rids of the "noise" and preserves the positions of the high-valued coefficients in the solution. In the StOMP algorithm, the $\sigma^{(k)}$ is the empirical noise variance and the threshold parameter ($\tau^{(k)}$) is chosen between 2 and 3. StOMP empirically showed that, for most sparse recovery problems, only three iterations are needed for convergence.

CoSamp

Another algorithm that has enjoyed a lot of success in sparse signal recovery is the CoSamp. The algorithm is simple to implement, but the logic behind its success is mathematically involved. There is no simple intuitive way to understand its working.

In CoSamp, positions of the top 2s coefficients of the correlate (c) are chosen as the intermediate support. The values at the support are updated accordingly; this is an intermediate variable b. The solution at the current iteration is updated by pruning b and keeping the top s coefficients in it.

COSAMP

Initialize: $\Lambda = \{\}$ and $x = 0$.
Until convergence repeat:
Compute: $c = A^T\left(y - Ax_k\right)$
Detect support: $l^{(k)} = \left\{j : |c_j| \in \text{top 2s coefficients in } |c_j|\right\}$
Append to existing support: $\Lambda = \Lambda \cup l^{(k)}$
Estimate intermediate: $b = \min_x \|y - A_\Lambda x_\Lambda\|_2^2$
Update: $x^{(k+1)} = \text{Top s values in b}$

To elucidate the algorithm, let us take the same example as in StOMP. We have three non-zero values at 2, 9, and 14. We do not show the original solution; it is the same as in Figure 2.7. After computing A^Ty, CoSamp selects the top 2s coefficients—in this case, six of them. This turns out to be positions 14, 2, 15, 4, 16, 12 (in that order). The intermediate variable b is computed, and the top 3 are preserved in the solution. This is shown in Figure 2.12.

In the second iteration, $r = y - Ax^{(0)}$ is computed, and the top six indices (9, 4, 1, 3, 16, 2) from A^Tr are added to the existing support (2, 12, 14); thus, the new support becomes (1, 2, 3, 4, 9, 12, 14, 16). The value of b is computed from this support, from which the top three (s) values are kept in the solution after the second iteration. This is shown in Figure 2.13.

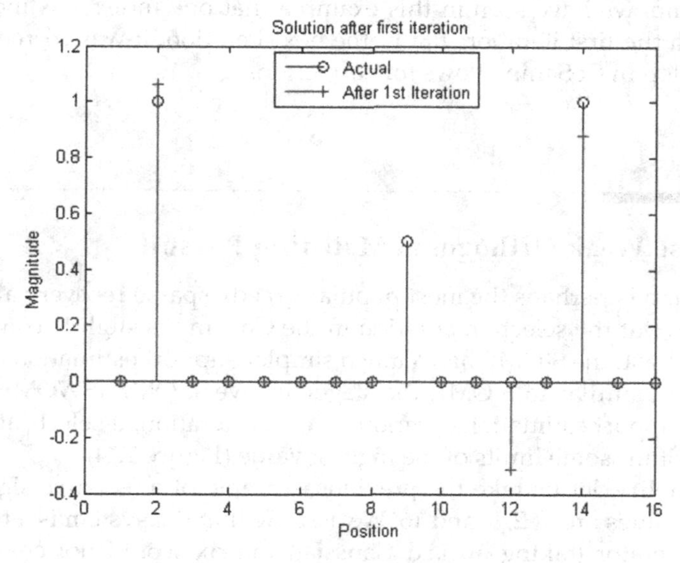

FIGURE 2.12
Solution after one iteration of CoSamp.

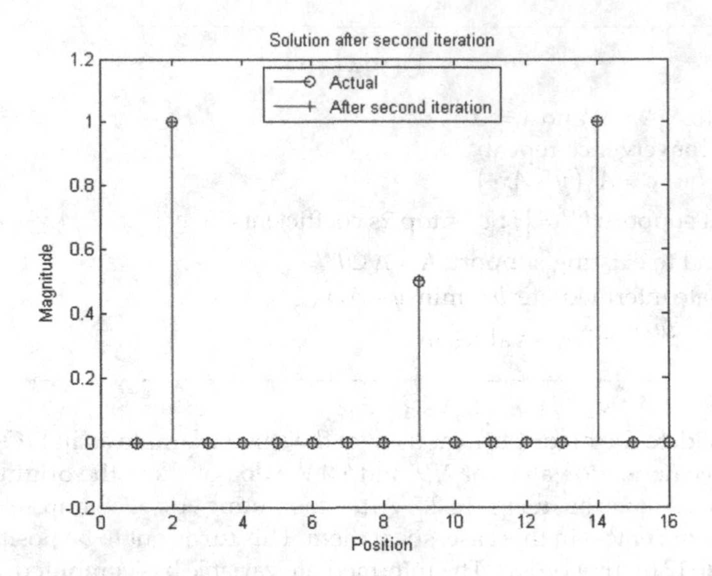

FIGURE 2.13
Solution after two iterations of CoSamp.

In this case, the solution is reached after the second iteration.

The advantage of the CoSamp over OMP and StOMP is that the CoSamp can rectify/prune an incorrectly selected index. In OMP or StOMP, if an index is incorrectly selected, it stays in the solution for future iterations. In the CoSamp, we have seen in this example that one index was incorrectly selected in the first iteration, but in the next iteration, it was corrected. The pruning step in CoSamp allows for correction.

Stagewise Weak Orthogonal Matching Pursuit

The CoSamp is perhaps the most popular greedy sparse recovery algorithm after OMP. But the selection criterion in the CoSamp is slightly complicated to say the least; the StOMP had a much simpler support estimation criterion. In a fashion similar to StOMP, the stagewise weak OMP (SWOMP) has an intuitive support estimation criterion. In every iteration, it selects all indices that are within some limits of the highest value (Figure 2.14).

To exemplify, let us take the previous example of a 3-sparse signal. The non-zero values are at 2, 9, and 16. We assume that the system is a restricted Fourier operator (taking an i.i.d Gaussian matrix would not change anything). As usual, we compute $A^T y$ and find out its maximum value (say m). To estimate the support, we find all positions that are within α (between 0 and 1) times the maximum value. If we use the value of $\alpha = 0.7$ (Figure 2.15),

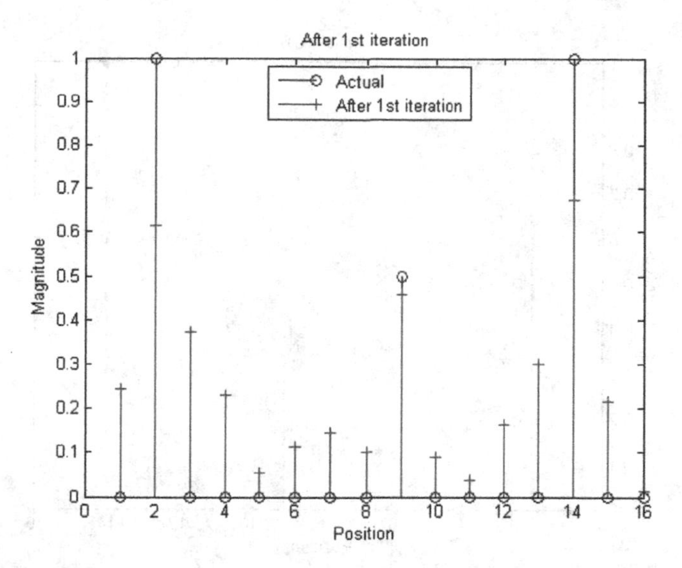

FIGURE 2.14
SWOMP after one iteration.

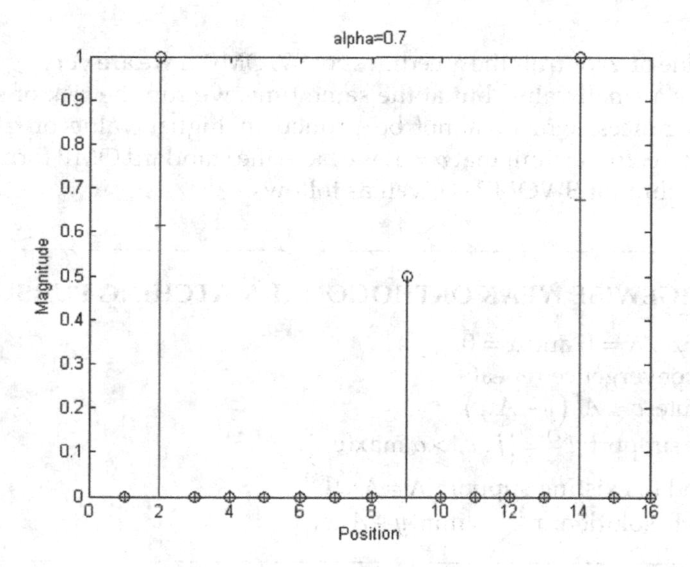

FIGURE 2.15
Support for $\alpha = 0.7$.

we will select two indices 2 and 14, but if we select $\alpha = 0.6$ (Figure 2.16), we will select all the three indices 2, 9, and 14.

Once the support is detected, finding the solution at that iteration proceeds as before, that is, via pseudoinverse. For the second case of $\alpha = 0.6$, the correct solution is obtained in the first iteration. For the case where $\alpha = 0.7$, one has to do a second iteration.

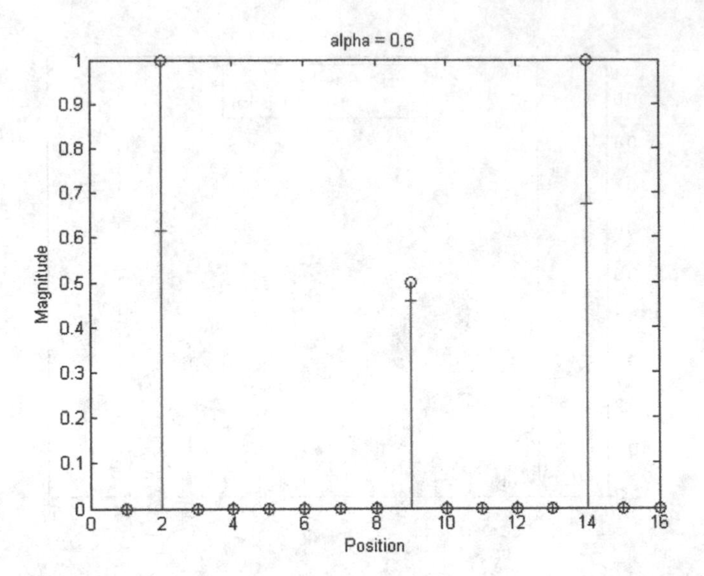

FIGURE 2.16
Support for $\alpha = 0.6$.

The value of α controls the greediness of SWOMP. If we are very aggressive, we choose a small value, but at the same time, we run the risk of selecting spurious indices, which cannot be pruned. A higher value of α is more conservative. In the limit that $\alpha = 1$, we have the standard OMP formulation. The algorithm for SWOMP is given as follows:

STAGEWISE WEAK ORTHOGONAL MATCHING PURSUIT

Initialize: $\Lambda = \{\}$ and $x = 0$.
Until convergence repeat:
Compute: $c = A^T\left(y - Ax_k\right)$
Detect support: $l^{(k)} = \left\{j : |c_j| > \alpha \max(c_j)\right\}$
Append to existing support: $\Lambda = \Lambda \cup l^{(k)}$
Estimate solution: $x^{(k+1)} = \min_x \left\|y - A_\Lambda x_\Lambda\right\|_2^2$

Gradient Pursuits

We mentioned earlier that OMP has the advantage of not repeatedly choosing the same index, unlike MP. The price we need to pay for this is the extra computational load of solving a least squares problem in every iteration.

This may become a computational bottleneck for large systems. In order to address the computational issue, gradient pursuit techniques have been proposed. The main idea here is that, instead of solving the exact least squares problem in every iteration, we take a step toward the solution along the gradient.

GRADIENT PURSUIT

Initialize: $\Lambda = \{\}$ and $x = 0$.
Until convergence repeat:
Compute: $c = A^T\left(y - Ax_k\right)$
Detect support: $l^{(k)} = \left\{j : \left|c_j\right| > \alpha \max(c_j)\right\}$
Append to existing support: $\Lambda = \Lambda \cup l^{(k)}$
Estimate solution: $x^{(k+1)} = x^{(k)} + ad^{(k)}$

Here, $d^{(k)}$ is the update direction, and a is the step size. The idea is simple and follows from basic gradient descent techniques. The step size can be chosen via various methods—steepest descent/scent, conjugate gradient, etc.

Here, we have shown it for an OMP-like selection criterion, but other greedier methods such as StOMP, CoSamp, and SWOMP can also be used. The only change is in the computation of the solution—instead of exactly orthogonalizing it by computing the least squares/pseudoinverse solution, a cheap gradient-based update is done to save computational time.

Suggested Reading

Y. C. Pati, R. Rezaiifar and P. S. Krishnaprasad, Orthogonal matching pursuit: Recursive function approximation with applications to wavelet decomposition, *Asilomar Conference on Signals, Systems and Computers*, Pacific Grove, CA, pp. 40–44, 1993.

J. A. Tropp and A. C. Gilbert, Signal recovery from random measurements via orthogonal matching pursuit, *IEEE Transactions on Information Theory*, 53 (12), 4655–4666, 2007.

D. L. Donoho, Y. Tsaig, I. Drori and J. L. Starck, Sparse solution of underdetermined systems of linear equations by stagewise orthogonal matching pursuit, *IEEE Transactions on Information Theory*, 58 (2), 1094–1121, 2012.

D. Needell and J. A. Tropp, CoSaMP: Iterative signal recovery from incomplete and inaccurate samples, *Applied and Computational Harmonic Analysis*, 26 (3), 301–321, 2008.

T. Blumensath and M. E. Davies, Stagewise weak gradient pursuits, *IEEE Transactions on Signal Processing*, 57 (11), 4333–4346, 2009.

Appendix: MATLAB® Codes

Ortogonal Matching Pursuit

```
% Orthogonal Matching Pursuit
% This function returns a column vector recovered using OMP
  algorithm.
% length(x)-->size(A,2);
% A--> measurement matrix
% y--> test vector: y=Ax
% k--> sparsity level of x

function x = OMP(A,y,k)
r=y;                        % initial Residue
O=[];                       % initialisation of Support Vectors
x=zeros(size(A,2),1);       % initialisation of x
for i=1:k
   c=A'*r;                  % Correlation
   [~,ind]=max(abs(c));     % Index with max correlation
   O=[O ind];               % Update support
   Ao=A(:,O');              % Updated measurement matrix
   x1=Ao\y;                 % min ||y-Ao*x||
   r=y-Ao*x1;               % ith step residual
end
x(O')=x1;
end
```

Stagewise Orthogonal Matching Pursuit

```
% Stagewise Orthogonal Matching Pursuit
% This function returns a column vector recovered using
  StOMP algorithm.
% length(x)=size(A,2);
% A--> measurement matrix
% y--> test vector: y=Ax
% N--> Number of iterations/stages

function x = StOMP(A,y,N)
r=y;                     % initial Residue
O=[];                    % initialisation of Support Vectors
n=size(A,2);
x=zeros(n,1);            % initialisation of x
t=3;                     % threshold parameter
sd=norm(r);
for i=1:N
```

```
    if A*x==y
        break
    end
    c=A'*r;                     % Correlation
    sd=norm(r)/sqrt(n);         % noise level(standard deviation)
    ind=find                    % find the desired indices greater
     (abs(c)>=t*sd);              than
threshold
    O=union(O,ind);             % Update Support
    Ao=A(:,O);                  % Updated measurement matrix
    x1=Ao\y;                    % min ||y-Ao*x||
    r=y-Ao*x1;                  % ith step residual
    x(O)=x1;
end
end
```

CoSamp

```
% Compressive Sampling Matching Pursuit
% This function returns a column vector recovered using
  CoSaMP algorithm.
% length(x)=size(A,2);
% A--> measurement matrix
% y--> test vector: y=Ax
% k--> sparsity level of x
% N--> number of Iterations

function x = CoSaMP(A,y,k,N)
r=y;                            % initial Residue
supp=[];                        % initialisation of Support
                                  Vectors
x=zeros(size(A,2),1);           % initialisation of x
% Repeat until convergence
% this is for fixed number
of iterations for i=1:N
    c=abs(A'*r);                % Correlation
    [~, O]=sort
    (c,'descend');
    O=O(1:2*k);                 % Select the top 2k support
    O=[supp;O];                 % Update support
    b=A(:,O)\y;                 % min ||y-Ao*x||
    [~, v]=sort(abs(b),
    'descend');
```

```
    ind=v(1:k);
    x1= b(ind);                    % upddate x by pruning b
                                   and taking top k
values
    supp=O(ind);                   % update support
    r=y-A(:,supp)*x1;              % ith step residual
end
x(supp)=x1;
end
```

Stagewise Weak Orthogonal Matching Pursuit

```
% Stagewise Weak Orthogonal Matching Pursuit
% This function returns a column vector recovered using
  SWOMP algorithm.
% length(x)=size(A,2);
% A--> measurement matrix
% y--> test vector: y=Ax
% k--> Sparsity level of x
% t--> parameter of threshold

function x = SWOMP(A,y,k,t)
r=y;                             % initial Residue
O=[];                            % initialisation of Support
                                   Vectors
x=zeros(size(A,2),1);            % initialisation of x
while(length(O)<k)
    c=A*r;                       % Correlation
    cmax=max(abs(c));            % max correlation value
    ind=find(abs(c)>=t*cmax)
    ;                            % find the desired indices
                                   greater than
threshold
    O=[O;ind];                   % Update Support
    Ao=A(:,O);                   % Updated measurement matrix
    x1=Ao\y;                     % min ||y-Ao*x||
    r=y-Ao*x1;                   % ith step residual
end
x(O)=x1;
end
```

3

Sparse Recovery

We are interested in solving an under-determined system of linear equations where the solution is known to be sparse.

$$y_{m\times 1} = A_{m\times n}x_{n\times 1}, m < n \tag{3.1}$$

In general, such an under-determined system has infinitely many solutions. This is because A has a non-empty null space, that is,

$$Ab = 0, \text{ when } b \neq 0 \tag{3.2}$$

Thus, one can start with any solution of (3.1) and keep adding vectors from the null space of A to generate infinitely many solutions.

In simpler terms, (3.1) has more free variables than the number of equations; therefore, one can impute any value to the free variables without hampering the relationship expressed in (3.1).

We are not interested in the generic problem of under-determined linear systems; instead, we look at a very specific problem—the case where we know that the solution is sparse. In such a case, it has been shown that the sparse solution is unique in most cases. This is good news. But how do we solve it? Rather, can we solve it?

Assume that the solution is s-sparse, that is, only x has s non-zero values and the rest $(n-s)$ are zeroes. An s-sparse vector has 2s degrees of freedom, that is, 2s unknowns—s unknown positions and the corresponding s unknown values. Therefore, if the number of equations (m) is larger than 2s, we can be hopeful of being able to solve it. Intuitively, when the amount of information available (m) is larger than the number of unknowns (2s), we feel that the problem should be solvable. But the question still remains, how to solve?

In the previous chapter on greedy algorithms, we learnt some greedy solution techniques. If you remember, these techniques were basically used to solve the l_0-norm minimization problem.

$$\min_x \|x\|_0 \leq s \text{ such that } y = Ax \tag{3.3}$$

The l_0-norm counts the number of non-zeroes in the vector. In this case, the number of non-zeroes is supposed to not exceed s.

We now know that the sparse solution is unique in most cases. Therefore, we do not expect to have a sparse solution with s non-zeroes and another one with some other number of non-zeroes. Therefore, instead of solving (3.3), we can solve the following; the uniqueness of the existence of a sparse solution guarantees equivalence.

$$\min_{x} \|x\|_0 \text{ such that } y = Ax \tag{3.4}$$

In the last chapter, we discussed why this is a combinatorial problem (non-deterministic polynomial (NP) hard); we will not repeat the discussion. In this chapter, we will learn about convex and non-convex relaxations of (3.4).

In order to reach a sparse solution, people have been using the l_1-norm, a convex relaxation of the NP hard l_0-norm, much before the advent of compressed sensing (CS). They were solving problems like,

$$\min_{x} \|x\|_1 \text{ such that } y = Ax \tag{3.5}$$

Such techniques were routinely used in seismic imaging and in sparse regression problems. Therefore, it was natural that researchers in CS proposed the same techniques for solving (3.1). The success of l_1-norm for sparse recovery was known since the 1970s, but the reason behind its success was not well understood. The mathematicians who pioneered CS were the first to prove the equivalence of l_0–l_1 minimization.

It is difficult to understand the mathematics behind the equivalence and the conditions under which the equivalence holds. In this chapter, we will not go into the mathematical intricacies. We will understand equivalence geometrically.

Equivalence of l_0-Norm and l_1-Norm Minimization

What would happen if we wanted to achieve a sparse solution by using l_2-norm minimization?

$$\min_{x} \|x\|_2 \text{ such that } y = Ax \tag{3.6}$$

This has a nice closed-form solution, but the solution would not be sparse. This can be understood from Figure 3.1. The l_2-norm is represented by a hypersphere (a circle in our case); the solution of (3.6) is the point where the system of equations (hyperplane in general, but a straight line for us) is tangential to the circle of l_2 ball. As can be seen, this point has non-zero values for both the axes. Thus, the solution is not sparse. This is why the simple minimum l_2-norm solution is not useful for achieving a sparse solution.

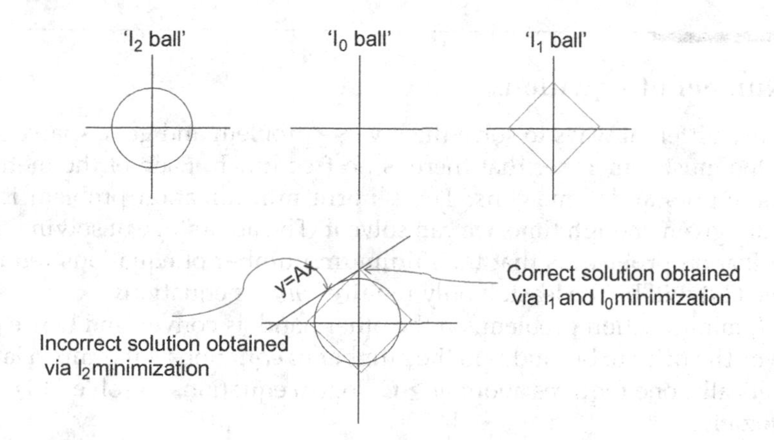

FIGURE 3.1
Equivalence of l_0–l_1 minimization.

Consider the ideal scenario of l_0 minimization. In this case, the l_0 ball consists of two axes. The l_0 ball intersects the line ($y = Ax$) at points where only one of the values is non-zero. In the simple two-dimensional (2D), case this is a sparse solution. We know that solving the l_0 minimization problem is not practical. Instead, we look for the minimum l_1-norm solution.

The l_1 ball is a square shown in Figure 3.1. The corners of the square lie on the axes. Thus, the l_1 ball touches the system of equations (straight line in our case) at one of its vertices, which are on the axes. Thus, the solution turns out to be sparse. The point where the l_0 ball and the l_1 ball touch the hyperplane (representing the system of equations) is the same. This elucidates the equivalence of l_0–l_1 norm geometrically.

In fact, there can also be a non-convex l_p-norm ($0 < p < 1$) minimization problem. This is depicted in Figure. 3.2. The l_p-norm is non-convex, but its vertices lie on the axes. Hence, the l_p ball touches the system of equations at the vertices and therefore yields the desired sparse solution.

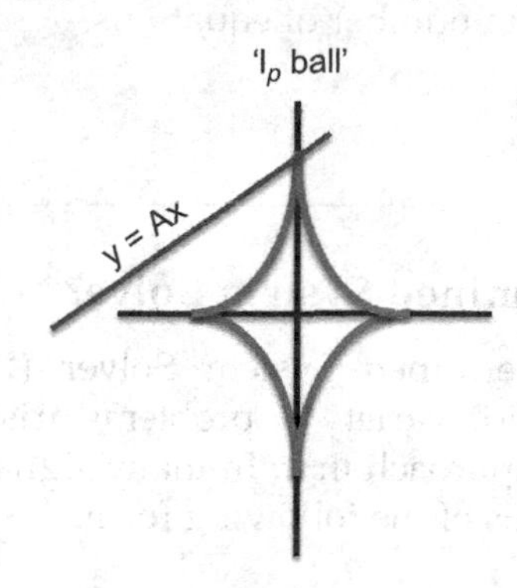

FIGURE 3.2
Geometry of l_p-norm minimization.

On Number of Equations

There are different ways to solve the inverse problem and get a sparse solution. One must remember that there is no free lunch. Each of the methods has some pros and some cons. The l_0-norm minimization problem is NP hard. But given enough time, we can solve it. The advantage of solving the l_0 minimization problem is that the minimum number of equations required to solve (3.1) will be the least. It only requires $m \geq 2s$ equations.

The l_1 minimization problem, on the other hand, is convex and hence easy to solve. The price to be paid is in the number of equations—it is much larger. Theoretically, one requires about $m \geq Cs \log(n)$ equations to solve (3.1) via l_1 minimization.

The l_p-norm lies between the l_0-norm and l_1-norm. It can be solved efficiently but is always fraught with the danger of being stuck in a local minimum, since the l_p-norm is non-convex. The number of equations required to solve (3.1) by l_p minimization is given by,

$$m \geq pC_1 s \log(n) + C_2 s \tag{3.7}$$

This is intermediate between l_1-norm and l_0-norm minimization. When the value of p is small (toward 0), the number of equations required increases linearly in s, as in the case of l_0 minimization. When p is high (nearing 1), the number of equations is determined by the first term of (3.7)—it is similar to the requirement of l_1 minimization.

You must be wondering why larger number of equations is a con. In most signal processing applications, the linear system (3.1) will be associated with some sampling process; more equations would mean more sampling, and there is always some cost associated with sampling—more power, more time, more harmful, etc. Our task would be to reduce the sampling requirement; in such a case, we would look for techniques that can solve the linear system with the minimum number of equations.

FOCally Underdetermined System Solver

The FOCally Underdetermined System Solver (FOCUSS) technique is generic. It can solve a wide variety of problems arising in sparse recovery. We discuss the generic approach first. In many signal processing problems, we need to solve problems of the following form.

$$\min_x E(x) \text{ subject to } y = Ax \tag{3.8}$$

where $E(x)$ is a diversity measure, for example, for sparse recovery, it is the l_p-norm $(0 < p \leq 1)$, and for low-rank recovery, it is the Schatten-q norm $(0 < q \leq 1)$.

The Lagrangian for (3.8) is,

$$L(x, \lambda) = E(x) + \lambda^T (Ax - b) \tag{3.9}$$

Following the theory of Lagrangian, the stationary point for (3.9) needs to be obtained by solving

$$\nabla_x L(\hat{x}, \hat{\lambda}) = 0 \tag{3.10}$$

$$\nabla_\lambda L(\hat{x}, \hat{\lambda}) = 0 \tag{3.11}$$

Solving (3.10) requires the gradient of $E(x)$ with respect to x. Please note that to preserve generality, the explicit functional form of $E(x)$ has not been stated. In the diversity measures of interest to us, the gradient of $E(x)$ can be expressed as,

$$\nabla_x E(x) = \alpha(x)\Pi(x)x \tag{3.12}$$

where $\alpha(x)$ is a scalar and $\Pi(x)$ is a diagonal matrix.

Solving (3.10) and applying (3.12) yields,

$$\hat{x} = -\frac{1}{\alpha(\hat{x})}\Pi(\hat{x})^{-1}A^T\hat{\lambda} \tag{3.13}$$

$$\hat{\lambda} = -\alpha(x)(A\Pi(\hat{x})^{-1}A^T)^{-1}y \tag{3.14}$$

Substituting (3.14) into (3.13) yields,

$$\hat{x} = \Pi(\hat{x})^{-1}A^T(A\Pi(\hat{x})^{-1}A^T)^{-1}y \tag{3.15}$$

Expression (3.15) is an implicit function of (x). Therefore, it needs to be solved iteratively. The kth iteration is given by:

$$x(k+1) = \Pi(x(k))^{-1} A^T (A\Pi(x(k))^{-1} A^T)^{-1} y \tag{3.16}$$

However, there is a basic problem with this FOCUSS approach. The theory of linear Lagrangian used to derive the FOCUSS algorithm is valid for convex problems, that is, for convex diversity measures. However, FOCUSS is almost always used for solving non-convex diversity measures such as l_p-norm minimization. There is no theoretical guarantee that FOCUSS will converge to the desired solution for such non-convex problems, but practical experiments show that this method (with a little modification) almost always provides exceptionally good results.

l_p Minimization Using FOCally Underdetermined System Solver

Using this general methodology, we will learn how to derive the algorithm for l_p-norm minimization. The problem we need solve is (3.17).

$$\min_x \|x\|_p^p \text{ subject to } y = Ax,\ \|x\|_p^p = \sum_{i=1}^{n} |x_i|^p \tag{3.17}$$

The diversity measure to be minimized is

$$\|x\|_p = \left(\sum_{i=1}^{n} |x_i|^p\right)$$

The gradient operator is defined as

$$\nabla_x = \left[\frac{\partial}{\partial x_1}, \ldots, \frac{\partial}{\partial x_n}\right]^T$$

The partial derivative with respect to x_i is,

$$\frac{\partial |x|^p}{\partial x_i} = p|x_i|^{p-2} x_i \tag{3.18}$$

Therefore, the gradient is,

$$\nabla_x \|x\|^p = p \cdot diag\left(|x_i|^{p-2}\right)x$$

Here, $diag\left(|x_i|^{p-2}\right)$ creates a diagonal matrix, with its diagonal elements as $|x_i|^{p-2}$. Comparing this with (3.12), we see that,

$$\alpha(x) = p$$

$$\Pi(x) = diag\left(|x_i|^{p-2}\right) \tag{3.19}$$

The original work on FOCUSS substituted the value of $\Pi(x)$ in (3.15) to solve the l_p-norm minimization problem. But l_p-norm minimization is a non-convex problem. Therefore, there is a chance that the optimization algorithm is stuck in a local minimum. To reduce the chances of converging to a local minimum, a simple modification has been proposed; this is given by,

$$\Pi(x) = diag\left(\left|x_i\right|^{p-2} + \delta\right) \tag{3.20}$$

where δ is the damping factor, whose value is reduced at each iteration.

The damping factor also sets bound to the values of the diagonal matrix and does not allow its inverse to become excessively large. The convergence (to a local minimum) of such a damped FOCUSS class of algorithms has been proven; it is shown that such algorithms have super-linear convergence.

Computing (3.15) at each iteration (k) is computationally expensive and numerically unstable, since it involves finding explicit matrix inverse. Looking closely at (3.15), we see that at each iteration, one needs to compute a weighted right inverse of A. This problem can be easily solved by some conjugate gradient solver. Thus, instead of computing (3.15) explicitly in each iteration, we break it into four substeps that can be solved efficiently.

$$R = diag\left(\left|x_i\right|^{p/2-1} + \delta\right)$$

$$\Phi = AR$$

compute $z(k+1)$ by solving $y = \Phi z$ via conjugate gradient (CG)

$$x(k+1) = Rz(k+1)$$

This solution precludes computing explicit inverses. Moreover, it does not require A to be specified explicitly and thus can be applied for fast operators. In practice, it is not necessary to solve the linear inverse problem perfectly, as only a few steps of conjugate gradient (CG) are sufficient.

Iterative Re-weighted Least Squares

This technique is very intuitive. Consider a generic weighted l_2 minimization problem.

$$\min_x \left\|W_x\right\|_2^2 \text{ such that } y = Ax \tag{3.21}$$

The unconstrained Lagrangian is given by,

$$L(x, \lambda) = x^T W^T W x + \lambda T (y - Ax) \tag{3.22}$$

Equating the gradient to zero, we get,

$$\nabla_x L = 2Wx + A^T \lambda = 0 \tag{3.23}$$

$$\nabla_\lambda L = Ax - y = 0 \tag{3.24}$$

From (3.23), we get,

$$x = -\frac{1}{2} W^{-1} A^T \lambda \tag{3.25}$$

assuming that W has an inverse.

Substituting x in (3.24) and solving for λ yields,

$$\lambda = -2(AW^{-1}A^T)^{-1} y \tag{3.26}$$

Substituting the value of λ back in (3.25) gives us the final closed-form solution.

$$x = W^{-1} A^T (AW^{-1} A^T)^{-1} y \tag{3.27}$$

We all know that the l_2 minimization (when W is identity) will not give us the desired sparse solution. We have to solve the l_p minimization problem. In iterative re-weighted least squares (IRLS), the l_p-norm is approximated by a weighted l_2-norm.

$$\begin{gathered} \|x\|_p^p = \|W_x\|_2^2 \\ w_i = |W_x|^{\frac{p}{2}-1}, W = diag(W_i) \end{gathered} \tag{3.28}$$

Since W is dependent on x, it is possible to formulate l_p minimization only in an iterative fashion. W is updated based on the values of x in the previous iteration. However, in that case, the relationship will not be exact, as the values of x are changing in every iteration; but when the solution converges, the weighted l_2-norm will be exactly representing the l_p-norm. Thus, in the asymptotic sense, the l_p minimization problem can be recast via IRLS as,

$$\min_x \|W_x\|_2^2 \text{ subject to } y = Ax \tag{3.29}$$

To ensure convergence, w_i is regularized as before by adding a small constant. Note the similarity between the IRLS solution and the FOCUSS solution.

Noisy (Practical) Scenario

In most practical problems, the system is corrupted by noise. Therefore, instead of having an ideal system as (3.1), we have to solve a noisy problem.

$$y = Ax + \eta, \eta \sim N(0, \sigma) \tag{3.30}$$

The noise is assumed to be normally distributed. Such a situation arises in almost all practical scenarios such as radar imaging, magnetic resonance imaging, and X-ray computed tompography. In such a noisy case, one needs to modify the original (equality constrained) problem and solve the following quadratic problem instead,

$$\min_x \|x\|_1 \text{ subject to } \|y - Ax\|_2^2 \leq \varepsilon,\ \varepsilon = m\sigma^2 \tag{3.31}$$

Such an l_1 minimization problem can arise in various flavors. The one in (3.31) is called basis pursuit denoising (BPDN); in signal processing, we mostly prefer this formulation because we have some information regarding the noise in the system.

In regression/machine learning, the alternate formulation is preferred.

$$\min_x \|y - Ax\|_2^2 \text{ subject to } \|x\|_1 \leq \tau \tag{3.32}$$

The data mismatch is minimized subject to the l_1-norm being lesser than some τ. The parameter τ is related to the sparsity. In regression problems, the noise parameter (σ) is not usually known, but the level of sparsity can be estimated. Thus, (3.32) is a more appropriate formulation. The sparse regression problem (3.32) is called Least Angle Selection and Shrinkage Operator (LASSO).

However, both BPDN and LASSO are constrained problems and hence are difficult to solve. In theory, there exists an equivalent unconstrained formulation; it is given by,

$$\min_x \|y - Ax\|_2^2 + \lambda \|x\|_1 \tag{3.33}$$

The three parameters λ, ε, and τ are related, but, in the general case, the relationship is not analytical, and hence, one cannot be estimated (by a formula) from the other. There is a trick called "cooling" to solve the constrained problem, given a solution to the unconstrained version (3.33).

Iterative Soft Thresholding Algorithm

The objective function (3.33) of our interest is not continuously differentiable, and there is no closed-form solution. The problem must be solved iteratively. We will follow the majorization-minimization (MM) approach to solve this problem. First, we provide an MM solution for minimizing the least squares. Later, we will show how the objective function/diversity measure is incorporated into it.

Figure 3.3 shows the geometrical interpretation behind the MM approach. The figure depicts the solution path for a simple scalar problem but essentially captures the MM idea.

Let $J(x)$ be the function to be minimized. Start with an initial point (at $k = 0$) x_k (Figure. 3.3a). A smooth function $G_k(x)$ is constructed through x_k, which has a higher value than $J(x)$ for all values of x, apart from x_k, at which the values are the same. This is the majorization step. The function $G_k(x)$ is constructed such that it is smooth and easy to minimize. At each step, minimize $G_k(x)$ to obtain the next iterate x_{k+1} (Figure 3.3b). A new $G_{k+1}(x)$ is constructed through x_{k+1}, which is now minimized to obtain the next iterate x_{k+2} (Figure 3.3c). As can be seen, the solution at every iteration gets closer to the actual solution.

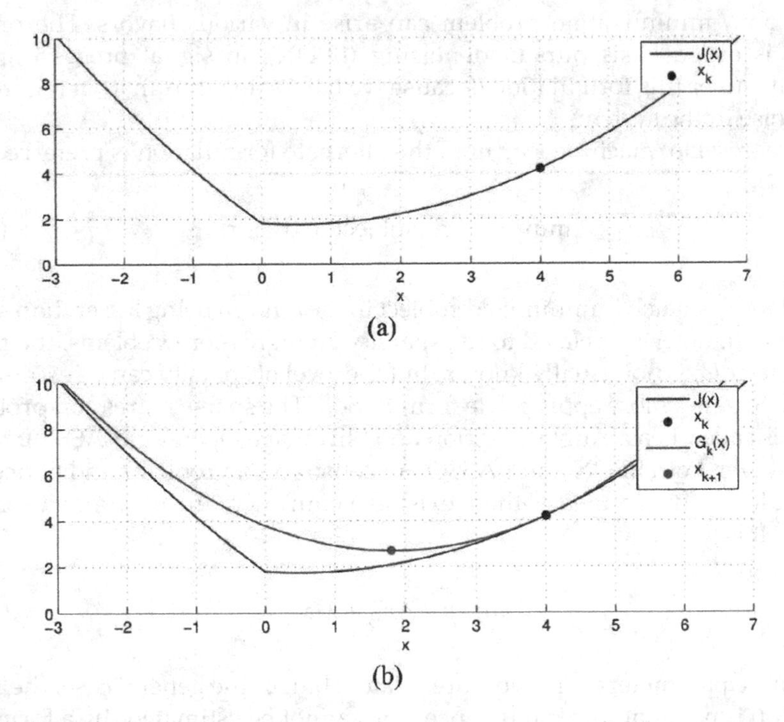

FIGURE 3.3
Majorization-minimization. (a) Function to be minimized. (b) Majorizer at x_k and minimum of majorizer at x_{k+1}. *(Continued)*

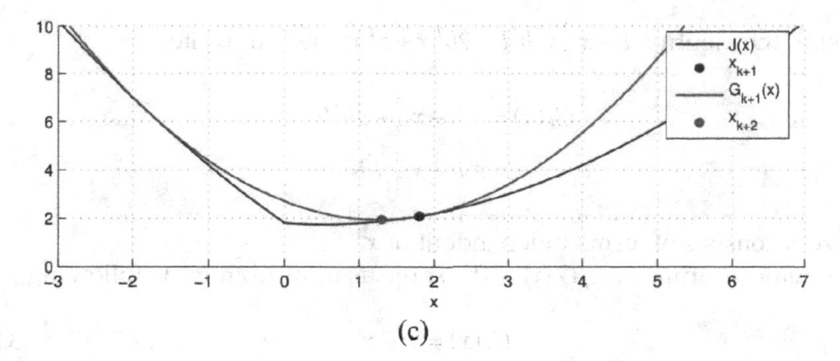

(c)

FIGURE 3.3 (Continued)
Majorization-minimization. (c) Majorizer at x_{k+1} and minimum of majorizer at x_{k+2}.

Majorization Minimization

First, let us consider the minimization of the least squares problem,

$$J(x) = \|y - Ax\|_2^2 \tag{3.34}$$

For this minimization problem, $G_k(x)$, the majorizer of $J(x)$, is chosen to be,

$$G_k(x) = \|y - Ax\|_2^2 + (x - x(k))^T \left(aI - A^T A\right)(x - x(k))$$

where a is the maximum eigenvalue of the matrix $A^T A$ and I is the identity.

One can check that at $x = x_k$, the expression $G_k(x)$ reduces to $J(x)$. At all other points, it is larger than $J(x)$; the value of "a" assures that the second term is positive definite.

$$\begin{aligned}
G_k(x) &= \|y - Ax\|_2^2 + (x - x(k))^T \left(aI - A^T A\right)(x - x(k)) \\
&= y^T y - 2y^T Ax + x^T A^T Ax + (x - x(k))^T \left(aI - A^T A\right)(x - x(k)) \\
&= y^T y + x(k)^T \left(aI - A^T A\right)x(k) - 2\left(y^T A + x(k)^T (aI - A^T A)\right)x(k) + ax^T x \\
&= a(-2b^T x - x^T x) + c
\end{aligned}$$

where $b = x(k) + \frac{1}{a} A^T \left(y - Ax(k)\right), c = y^T y + x(k)^T (aI - A^T A)x(k)$

Using the identity $\|b-x\|_2^2 = b^T b - 2b^T x + x^T x$, one can write,

$$G_k(x) = a\|b-x\|_2^2 - ab^T b + c$$
$$= a\|b-x\|_2^2 + K$$

where K consists of terms independent of x.

Therefore, minimizing $G_k(x)$ is the same as minimizing the following,

$$G_k'(x) = \|b-x\|_2^2 \tag{3.35}$$

where $b = x(k) + \frac{1}{a}A^T\left(y - Ax(k)\right)$.

This update is known as the Landweber iteration.

In all cases, our goal is to solve constrained optimization problems, where the objective function/diversity measure incorporates prior information about the solution and the constraint is the least squares data mismatch. The problem looks like:

$$\min_x \|x\|_1 \text{ such that } \|y - Ax\|_2^2 \leq \varepsilon \tag{3.36}$$

In theory, it can be alternately expressed as an equivalent unconstrained Lagrangian,

$$\min_x \|y - Ax\|_2^2 + \lambda\|x\|_1 \tag{3.37}$$

Theoretically, the constrained form (3.36) and the unconstrained form (3.37) are equivalent for correct choice of λ and ε. Unfortunately, for all practical problems, it is not possible to find the relation between the two explicitly. However, owing to the smoothness of the Pareto curve, the unconstrained problems for a decreasing sequence of λ are guaranteed to reach the solution desired by the constrained form. Based on this idea, a cooling algorithm can be devised to solve the constrained problem (3.36) by iteratively solving the unconstrained problems (3.37) with decreasing values of λ. The generic cooling algorithm is as follows:

1. Choose a high value of λ initially.
2. Solve the unconstrained optimization (3.37) for the given value of λ.
3. Decrease the value of λ and go back to Step 1.
4. Continue Steps 2 and 3, until the l_2-norm data mismatch $\left(\|y - Ax\|_2^2\right)$ between the data and the solution is less than ε.

Solving the Unconstrained Problem

Our goal is to solve,

$$\min_{x} \|y - Ax\|_2^2 + \lambda \|x\|_1 \tag{3.38}$$

The MM of (3.59) leads to combining the Landweber iterations (3.35) with l_1-norm regularization in (3.35), leading to the following problem,

$$G_k'(x) = \|b - x\|_2^2 + \frac{\lambda}{a}\|x\|_1 \tag{3.39}$$

Function (3.39) is actually de-coupled, that is,

$$G_k'(x) = \sum_i (b_i - x_i)^2 + \frac{\lambda}{a}|x_i| \tag{3.40}$$

Therefore, (3.40) can be minimized term by term, that is,

$$\frac{\partial}{\partial x_i} G_k'(x) = 2b_i - 2x_i + \frac{\lambda}{a} signum(x_i) \tag{3.41}$$

Setting the partial derivatives $\frac{\partial}{\partial x_i} G_k'(x)$ and solving for x give the graph shown in Figure. 3.4 with threshold $\lambda/2a$. That is, the minimizer of $G_k'(x)$

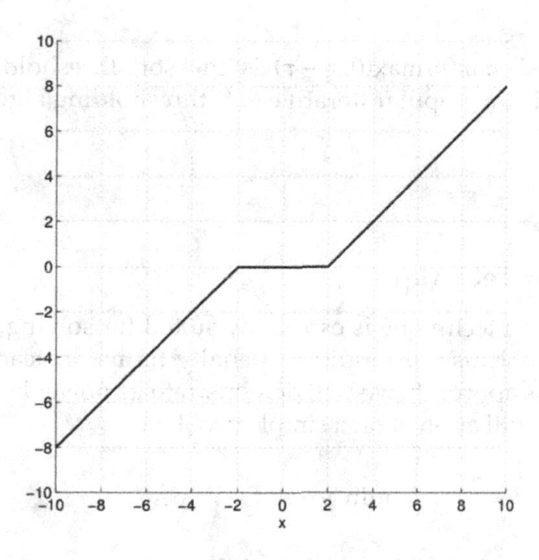

FIGURE 3.4
Soft threshold rule (with $\tau = 2$).

is obtained by applying the soft threshold rule to b with threshold $\lambda/2a$. The soft threshold rule is a non-linear function defined as,

$$soft(x,\tau)=\begin{cases} x+\tau & x<-\tau \\ 0 & |x|=\tau \\ x-\tau & x>\tau \end{cases} \tag{3.42}$$

Or more compactly,

$$x_i = signum(b_i)\max\left(0,|b_i|-\frac{\lambda}{2a}\right) \tag{3.43}$$

This leads to a simple two-step iterative solution for the unconstrained l_1 minimization algorithm:

Initialize: $x(0)=0$
Repeat until: $\|y-Ax\|_2^2 \le \varepsilon$
Step 1. $b = x(k)+\frac{1}{a}A^T\left(y-Ax(k)\right)$
Step 2. $x(k+1) = signum(b)\max\left(0,|b|-\frac{\lambda}{2a}\right)$
End

The operation $signum(v)\max(0,|v|-\tau)$ is the soft thresholding operation. This algorithm is the popular iterative soft thresholding algorithm (ISTA).

Split Bregman Technique

The Split Bregman technique is especially suited for solving unconstrained optimization problems with multiple penalty terms. Instead of going into the theory of this approach, we will demonstrate its power by illustration. In this chapter, we will apply it on a simple problem.

$$\min_x \|y-Ax\|_2^2 + \lambda\|x\|_1$$

Solving such a simple problem with Split Bregman is an overkill; nevertheless, it will help the readers understand the basic approach.

Let us substitute the variable in the penalty by $p = x$; with this substitution, the problem is converted to a constrained optimization problem of the form:

$$\min_{p,x} \|y - Ax\|_2^2 + \lambda \|p\|_1 \text{ such that } p = x \tag{3.44}$$

The Lagrangian for the above is:

$$L(x,p,\mu) = \|y - Ax\|_2^2 + \lambda \|p\|_1 + \mu^T (p - x) \tag{3.45}$$

The Lagrangian imposes the strict constraint of equality between the variables p and x in every iteration. To relax this constraint, the augmented Lagrangian formulation is used,

$$AL(x,p) = \|y - Ax\|_2^2 + \lambda \|p\|_1 + \mu \|p - x\|_2^2 \tag{3.46}$$

This formulation does not enforce strict equality. The main idea behind the augmented Lagrangian is to relax the equality in the initial iterations but enforce it toward the end (while the solution converges). In order to do so, the value of μ is varied; usually, it is kept low initially, but as the solution converges, its value is progressively increased.

The Split Bregman approach achieves the same end but by a different approach. It introduces a proxy relaxation variable that is automatically updated in every iteration, thereby foregoing the requirement of heating/increasing μ. Thus, the Split Bregman formulation is:

$$SB(x,p) = \|y - Ax\|_2^2 + \lambda \|p\|_1 + \mu \|p - x - b\|_2^2 \tag{3.47}$$

where b is the Bregman relaxation variable.

How do we solve it? One can directly solve (3.47) by a gradient descent approach, but such a solution is not elegant. One can easily observe that the variables x and p are already separable, and hence, alternating direction method of multipliers (ADMM) can be invoked to solve (3.47) via two subproblems:

$$\text{P1}: \min_{x} \|y - Ax\|_2^2 + \mu \|p - x - b\|_2^2 \tag{3.48}$$

$$\text{P2}: \min_{x} \lambda \|p\|_1 + \mu \|p - x - b\|_2^2 \tag{3.49}$$

Solving P1 is easy, since it is a least squares problem (see below)

$$\text{P1}: \min_{x} \left\| \begin{pmatrix} y \\ p - b \end{pmatrix} - \begin{pmatrix} A \\ I \end{pmatrix} x \right\|_2^2$$

Depending on the nature of the problem, one can invoke a suitable solver for solving it. If A is available as an operator, one can use any CG solver; in case A is not an operator but an explicit matrix (usually this is the case in tomographic problems), one can use randomized Kaczmarz method.

It is easy to observe that P2 is just an l_1-norm denoising problem. We just learnt that the solution for this is the soft thresholding; we have already shown the derivation form in the previous section.

$$p = signum(x+b)\max\left(0, |x+b| - \frac{\lambda}{2\mu a}\right) \tag{3.50}$$

The final step is to update the Bregman relaxation variable; this is done by a simple gradient update:

$$b \leftarrow p - x - b \tag{3.51}$$

Putting it all together, the Split Bregman algorithm for solving the l_1 minimization problem is:

Initialize: p and b
In every iteration:

Solve P1: $x = \min_x \left\| \begin{pmatrix} y \\ p-b \end{pmatrix} - \begin{pmatrix} A \\ I \end{pmatrix} x \right\|_2^2$

Solve P2: $p = signum(x+b)\max\left(0, |x+b| - \frac{\lambda}{2\mu a}\right)$

Update: $b \leftarrow p - x - b$

Solving l_0-Norm Minimization

We know that solving the l_0 minimization problem is NP hard. We discussed this in the previous chapter as well as in this one. However, researchers have proposed several approximate approaches to solve this problem. We will discuss popular ones: smoothed l_0 minimization and iterative hard thresholding algorithm (IHTA).

Smooth l_0-Norm Minimization

The smooth l_0-norm (SL0) minimization is proposed as a solution to the NP hard sparse estimation problem.

$$\min_x \|x\|_0 \text{ such that } y = Ax$$

It replaced the NP hard (spiky) l_0-norm by a parameterized smooth function whose smoothness could be varied. The smooth function allowed for the usage of gradient-based optimization techniques. The solution was achieved iteratively by gradually reducing the smoothness of the function.

$$\text{Define: } \gamma(x_j) = \begin{cases} 1 \text{ when } |x_j| > 0 \\ 0 \text{ when } |x_j| = 0 \end{cases}, j = 1 \ldots n$$

Based on the above definition, one can express the l_0-norm as follows,

$$\|x\|_0 = \sum_{j=1}^{n} \gamma(x_j) \tag{3.52}$$

SL0 replaces the non-smooth (spiky) function $\gamma(x)$ by a smooth zero-mean Gaussian function whose smoothness can be varied.

$$\text{Define: } f_\sigma(x_j) = e^{\left(-x_j^2/2\sigma^2\right)}, j = 1 \ldots n$$

The function is smooth when σ is large and becomes spiky when its value reduces. In the limit that the σ is zero, the above function has the following property,

$$\lim_{\sigma \to 0} f_\sigma(x_j) = \begin{cases} 1 \text{ when } |x_j| = 0 \\ 0 \text{ when } |x_j| > 0 \end{cases} \tag{3.53}$$

Therefore, $\lim_{\sigma \to 0} f_\sigma(x_j) = 1 - \gamma(x_j)$. This allows for approximating the $\|x\|_0$ by,

$$\lim_{\sigma \to 0} F_\sigma(x) = \sum_{j=1}^{n} f_\sigma(x_j) = \sum_{j=1}^{n} 1 - \gamma(x_j) = N - \|x\|_0$$

Therefore, the l_0-norm minimization problem can be recast as,

$$\max \lim_{\sigma \to 0} F_\sigma(x) \text{ subject to } y = Ax \tag{3.54}$$

Since the objective function is smooth, it is easy to solve (3.54) by gradient-based methods. The main idea behind the algorithm is that, at each iteration, (3.54) is solved for a particular value of σ; in the following iteration, the value of σ is decreased and (3.54) is solved again. This continues till the solution has converged.

ALGORITHM FOR SMOOTH l_0-NORM MINIMIZATION

Initialization – Obtain the initial solution $\hat{x}^{(0)} = \min \|b - Ax\|_2^2$.

At Iteration k – Continue the following steps till solution is reached (i.e., till σ is greater than a specified value)

1. Choose $\sigma = c \cdot \max\left(\|x_i\|_2\right)$, where c is a constant greater than 4.
2. Maximize (3.22) for the current value of σ. Steepest ascent method is used to achieve this.
   a. Initialize, $s = x^{(k+1)}$
   b. Let $\Delta s = \left[s_1 \cdot e^{-s_1^2/2\sigma^2}, \ldots, s_n \cdot e^{-s_n^2/2\sigma^2}\right]^T$
   c. Update: $s \leftarrow s - \mu \Delta s$, where μ is a small constant
   d. Project the solution back to the feasible set by $s \leftarrow s - A^T(AA^T)^{-1}(As - x(k-1))$. To avoid calculating explicit inverses, the projection is broken down into the following three steps:
      i. Compute: $t_1 = As - x(k-1)$.
      ii. Solve: $\hat{t}_2 = \min \|t_1 - At_2\|_2$.
      iii. Project: $s \leftarrow s - \hat{t}_2$.

Set $x^{(k)} = s$ and return to Step 1 until convergence.

Iterative Hard Thresholding Algorithm

The problem remains the same as before—that of solving an l_0-norm minimization problem. But now, we are interested in solving the practical problem, where the system is corrupted by additive Gaussian noise. In that case, we would like to solve:

$$\min_x \|x\|_0 \text{ subject to } \|y - Ax\|_2^2 \leq \varepsilon \tag{3.55}$$

As before, we will first derive an algorithm to solve the unconstrained version of (3.55).

$$\min_x \|y - Ax\|_2^2 + \lambda \|x\|_0 \tag{3.56}$$

In deriving the algorithm for ISTA, we followed the MM technique. We follow exactly the same approach here.

With Landweber iterations, (3.56) can be expressed as follows,

$$\min_x G_k'(x) = \|b - x\|_2^2 + \frac{\lambda}{a}\|x\|_0 \tag{3.57}$$

where $b = x(k) + \frac{1}{a}A^T\left(y - Ax(k)\right)$

This is a decoupled problem and can be expressed as,

$$G_k'(s) = \left(b_1 - x_1\right)^2 + \frac{\lambda}{a}|x_1|^0 + \ldots + \left(b_n - x_n\right)^2 + \frac{\lambda}{a}|x_n|^0 \tag{3.58}$$

We can process (3.58) element-wise.

To derive the minimum, two cases need to be considered: case 1 – $x^{(i)} = 0$ and case 2 – $x^{(i)} \neq 0$. The element-wise cost is 0 in the first case. For the second case, the minimum is reached when $x(i) = b(i)$.

Comparing the cost in both cases,

$$0 \text{ if } x_i = 0$$

$$-\left(b_i\right)^2 + \frac{\lambda}{a} \text{ if } x_i = b_i$$

This suggests the following updates rule,

$$x^{(i)} = \begin{cases} b^{(i)} \text{ when } \left|b^{(i)}\right| > \lambda / 2a \\ 0 \text{ when } \left|b^{(i)}\right| \leq \lambda / 2a \end{cases} \tag{3.59}$$

This is the iterative hard thresholding algorithm. In a succinct fashion, it is just a two-step algorithm like ISTA.

Initialize: $x(0) = 0$
Repeat until: $\|y - Ax\|_2^2 \leq \varepsilon$
Step 1. $b = x(k) + \frac{1}{a} A^T \left(y - Ax(k)\right)$
Step 2. $x(k+1) = HardThreshold\left(|b|, \frac{\lambda}{2a}\right)$
End

l_p-Norm Minimization—Noisy Scenario Iterative Re-weighted Least Squares

We have studied the FOCUSS- and IRLS-based technique for l_p minimization. Those techniques were especially suited for solving the noise-free scenario. Those formulations can also be used to solve for the noisy case. In that case, the l_p-norm minimization is framed as,

$$\min_x \|y - Ax\|_2^2 + \lambda \|x\|_p^p \tag{3.60}$$

Using the IRLS technique, the l_p-norm penalty can be expressed as a weighted l_2-norm.

$$\min_x \|y - Ax\|_2^2 + \lambda \|Wx\|_2^2 \tag{3.61}$$

Solving (3.61) is straightforward, since it is just a least squares problem,

$$\min_x \left\| \begin{pmatrix} y \\ 0 \end{pmatrix} - \begin{pmatrix} A \\ \sqrt{\lambda W} \end{pmatrix} x \right\|_2^2 \tag{3.62}$$

This can be solved by using CG or any other technique.

In every iteration, the weight W is computed based on the values of x from the previous iteration. This is the same weight matrix as in (3.28). The full algorithm is given succinctly.

Initialize: $W = \text{Identity}; x(0) = \min_x \|y - Ax\|_2^2 + \lambda \|x\|_2^2$
Until convergence repeat:

Compute $W - w_i = |x_i|^{\frac{p}{2}-1}, W = diag(w_i + \varepsilon)$

Update $x - x(k) = \min_x \|y - Ax\|_2^2 + \lambda \|Wx\|_2^2$

Modified Iterative Soft Thresholding Algorithm

There is another solution to the l_p-norm minimization problem based on the MM approach. The objective remains the same—that of solving (3.60). After Landweber update, this can be expressed as,

$$\min_x G_k'(x) = \|b - x\|_2^2 + \frac{\lambda}{a}\|x\|_p^p \tag{3.63}$$

where $b = x(k) + \frac{1}{a}A^T(y - Ax(k))$

As before, the expression (3.63) is decoupled,

$$G_k'(x) = \sum_i (b_i - x_i)^2 + \frac{\lambda}{a}|x_i|^p \tag{3.64}$$

Therefore, (3.64) can be minimized term by term. Its gradient is expressed as,

$$\frac{\partial G_k'(x)}{\partial x_i} = 2b_i - 2x_i + \frac{\lambda}{a} p |x_i|^{p-1} signum(x_i) \tag{3.65}$$

We have expressed the derivative in this form so as to have an expression at par with ISTA (3.41); we repeat it for the sake of convenience.

$$\frac{\partial}{\partial x_i} G_k'(x) = 2b_i - 2x_i + \frac{\lambda}{a} signum(x_i) \tag{3.66}$$

Comparing (3.65) to (3.66), we see that (3.65) can also be efficiently updated by using soft thresholding, albeit with a different threshold $-\frac{\lambda}{2a} p |x_i|^{p-1}$. Here, the threshold becomes dependent on x. The problem can be easily mitigated by computing the threshold based on the value of x from the previous iteration. This is a modified ISTA

Initialize: $x(0) = 0$

Repeat until: $\|y - Ax\|_2^2 \leq \varepsilon$

Step 1. $b = x(k) + \frac{1}{a} A^T \left(y - Ax(k)\right)$

Step 2. $x(k+1) = signum(b) \max\left(0, |b| - \frac{\lambda}{2\alpha} p |x_i|^{p-1}\right)$

End

Suggested Reading

B. D. Rao, and K. Kreutz-Delgado, An affine scaling methodology for best basis selection, *IEEE Transactions on Signal Processing*, 47 (1), 187–200, 1999.

R. Chartrand and W. Yin, Iteratively reweighted algorithms for compressive sensing, *IEEE ICASSP*, 2008.

I. W. Selesnick and M. A. T. Figueiredo, Signal restoration with overcomplete wavelet transforms: Comparison of analysis and synthesis priors, *Proceedings of SPIE*, 7446 (Wavelets XIII), 2009.

T. Blumensath and M. E. Davies, Iterative thresholding for sparse approximations, *Journal of Fourier Analysis Applications*, 14 (5), 629–654, 2008.

A. Majumdar and R. K. Ward, On the choice of compressed sensing priors: An experimental study, *Signal Processing: Image Communication*, 27 (9), 1035–1048, 2012.

I. Selesnick, Sparse Signal Restoration, CNX. Available at http://cnx.org/contents/c9c730be-10b7- 4d19-b1be-22f77682c902@3/Sparse-Signal-Restoration.

Appendix: MATLAB® Codes

Cooled ISTA

```
function beta = CooledIST(H, Ht, y, err, normfac,
insweep, tol, decfac)

% alpha = maxeig(H'H)
% for restricted fourier type of H, alpha > 1;

if nargin < 4
    err = 1e-6;
end
if nargin < 5
    normfac = 1;
end
if nargin < 6
    insweep = 200;
end
if nargin < 7
    tol = 1e-4;
end
if nargin < 8
    decfac = 0.5;
end

alpha = 1.1*normfac;
x = zeros(length(Ht(y)),1);
lambdaInit = decfac*max(abs(Ht(y))); lambda = lambdaInit;
f_current = norm(y-H(x_initial)) + lambdaInit*norm(x,1);

while lambda > lambdaInit*tol
    % debug lambda
    for ins = 1:insweep
        f_previous = f_current;
        Hx = H(x);
        x = SoftTh(x + (Ht(y - Hx))/alpha, lambda/(2*alpha));
        f_current = norm(y-H(x)) + lambda*norm(x,1);
        if norm(f_current-f_previous)/norm(f_current +
        f_previous)<tol
            break;
        end
    end
    norm(y-H(x))
    if norm(y-H(x))<err
        break;
    end
```

```
    lambda = decfac*lambda;
end
beta = x;

    function z = SoftTh(s,thld)
        z = sign(s).*max(0,abs(s)-thld);
    end
end
```

Cooled IHTA

```
function beta = CooledIHT(H, Ht, y, err, x_initial, normfac,
insweep, tol, decfac)

% alpha = maxeig(H'H)
% for restricted fourier type of H, alpha > 1;

if nargin < 4
    err = 1e-6;
end
if nargin < 5
    x_initial = zeros(length(Ht(y)),1);
end
if nargin < 6
    normfac = 1;
end
if nargin < 7
    insweep = 200;
end
if nargin < 8
    tol = 1e-4;
end
if nargin < 9
    decfac = 0.5;
end

alpha = 1.1*normfac;
x = x_initial;
lambdaInit = decfac*max(abs(Ht(y)));lambda = lambdaInit;
f_current = norm(y-H(x_initial)) + lambdaInit*norm(x,1);

while lambda > lambdaInit*tol
    for ins = 1:insweep
        f_previous = f_current;
        Hx = H(x);
        x = HardTh(x + (Ht(y - Hx))/alpha, lambda/
        (2*alpha));
        f_current = norm(y-H(x)) + lambda*norm(x,1);
        if norm(f_current-f_previous)/norm(f_current +
        f_previous)<tol
            break;
```

```
        end
    end
    if norm(y-H(x))<err
        break;
    end
    lambda = decfac*lambda;
end
beta = x;

    function z = HardTh(s,thld)
        z = (abs(s) >= thld) .* s;
    end
end
```

IRLS

```
function x = IRLS(A,At,y,p,iter)
%
% INPUT
%   A, At       : mxn matrix or operator;
%   y           : vector of dimension m; result.
%   p           : norm order; p = 1 default
%   iter        : maximum number of iterations (default = 50)
% OUTPUT
%   x           : vector of vector n; initial unknown
%
if nargin < 4
    p = 1;
end
if nargin < 5
    iter = 50;
end
% Set LSQR parameters
OptTol = 1e-6; epsilon = 1;
u_old = lsqr(@lsqrA,y,OptTol,100); j=0;
while (epsilon > 10^(-8)) && (j < iter)
    j = j + 1;
    w = (abs(u_old) + epsilon).^(1-p/2);
    t = lsqr(@lsqrAW,y,OptTol,100,[],[],(1./w).*u_old);
    u_new = w.*t;
    if lt(norm(u_new - u_old,2),epsilon^(1/2)/100)
        epsilon = epsilon /10;
    end
    u_old = u_new;
end
x=u_new;
```

```
%%%%%%%%%%%%%%%%%%%%%%%%%%%%%%%%%%%%%%%%%%%%%%%%%%%%%%%%%%%%
%%%%%%%%%%%%%
    function y = lsqrA(x,transpose)
        switch transpose
           case 'transp'
               y = At(x);
           case 'notransp'
               y = A(x);
        end
    end
    function y = lsqrAW(x,transpose)
        switch transpose
           case 'transp'
               y = w.*(At(x));
           case 'notransp'
               y = A(w.*x);
        end
    end
end
```

Modified ISTA for lp-minimization

```
function beta = LpSynthesis(H, Ht, y, p, normfac, err,
insweep, tol, decfac)
% Algorithm for solving problems of the form:
% min ||x||_p s.t. ||y-Hx||_2 < err
if nargin < 4
    p = 1;
end
if nargin < 5
    normfac = 1;
end
if nargin < 6
    err = 1e-6;
end
if nargin < 7
    insweep = 50;
end
if nargin < 8
    tol = 1e-4;
end
if nargin < 9
    decfac = 0.5;

end

x_initial = Ht(y);
alpha = 1.1*normfac;
x = x_initial;
```

```
lambdaInit = decfac*max(abs(Ht(y))); lambda = lambdaInit;
f_current = norm(y-H(x_initial)) + lambda*norm(x,p);

while lambda > lambdaInit*tol
    w = ones(length(x),1);
    for ins = 1:insweep
        f_previous = f_current;
        x = SoftTh(x + (Ht(y - H(x)))/alpha, w.*lambda/
        (2*alpha)); w = abs(x).^(p-1);
        f_current = norm(y-H(x)) + lambda*norm(x,p);

        if norm(f_current-f_previous)/norm(f_current +
        f_previous)<tol
            break;
        end
    end
%    norm(y-H(x))
   if norm(y-H(x))<err
       break;
    end
    lambda = decfac*lambda;
end
beta = x;

    function z = SoftTh(s,thld)
        z = sign(s).*max(0,abs(s)-thld);
    end
end
```

4

Co-sparse Recovery

We want to solve an under-determined linear inverse problem where the solution is known to be sparse. This is expressed as,

$$y = Ax + n \tag{4.1}$$

Here, we have assumed practical conditions where the system of equation may be noisy.

So far, we have studied the solution where the solution itself is sparse. But practically speaking, there is hardly an example where the solution will be so. Most natural signals are not sparse, for example, audio or biomedical signals have values everywhere, and so do images. But they do have a sparse representation in some other domain, for example, wavelet and discrete cosine transform (DCT). Such transforms follow the analysis synthesis equations:

$$\text{Analysis} : \alpha = Sx$$

$$\text{Synthesis} : x = S^T \alpha$$

Here, x is the signal, S is the sparsifying transform, and α represents the sparse coefficients. The analysis–synthesis equations hold for orthogonal[1] and tight-frame[2] systems. The synthesis equation allows (4.1) to be expressed in the familiar form:

$$y = AS^T \alpha + n \tag{4.2}$$

This is called the synthesis prior formulation. We have learnt both the greedy and optimal techniques for solving (4.2). We express the solution in the following form:

$$\min_{\alpha} \|\alpha\|_1 \text{ such that } \|y - AS^T \alpha\|_2^2 \le \varepsilon \tag{4.3}$$

---

[1] *Orthogonal*: $S^T S = I = SS^T$

[2] *Orthogonal*: $S^T S = I \neq SS^T$

Once the sparse coefficients are obtained, the signal can be reconstructed from the synthesis equation. Perhaps this is the most popular approach toward compressed sensing.

Note that the synthesis formulation is restrictive. Unless the sparsifying transform is orthogonal or tight-frame, this form (4.2) is not applicable. It restricts a large class of powerful linear transforms, such as the Gabor for biomedical signals or finite differencing for images. These transforms are known to have sparse outputs, but they do not follow the analysis–synthesis equations. But, one can alternately pose recovery in the analysis prior form:

$$\min_x \|Sx\|_1 \text{ such that } \|y - Ax\|_2^2 \leq \varepsilon \tag{4.4}$$

This is the co-sparse analysis prior formulation. Instead of trying to find a sparse solution, we find a solution that is co-sparse, that is, sparse in the transform domain. As one can see, we directly solve for the signal in this formulation.

It is easy to verify that for orthogonal systems, the synthesis and analysis forms are exactly the same, but these are not the same for tight-frames systems.

In this chapter, we will learn different techniques to solve for the co-sparse analysis prior formulation.

Majorization Minimization

Here, we learn to solve the generic l_p-minimization problem, where p can be anything between 0 and 1. We have learnt to solve problems by using the majorization-minimization (MM) approach before. The general approach is to solve the unconstrained formulation and then use cooling to solve the constrained version. The unconstrained version is expressed as,

$$\min_x \|y - Ax\|_2^2 + \lambda \|Sx\|_p^p \tag{4.5}$$

After standard MM, we get the following form,

$$G_k(x) = \|b - x\|_2^2 + \frac{\lambda}{a} \|Sx\|_p^p \tag{4.6}$$

where $b = x_k + \frac{1}{a} A^T (y - Ax_k)$; here, a is the maximum Eigenvalue of $A^T A$.

Unlike the synthesis prior formulation which was separable, the solution to the analysis prior form is not. Therefore minimizing it is slightly more involved. We take its gradient,

$$\nabla G_k(x) = 2b - 2x + \frac{\lambda}{a} S^T DSx$$

where $D = diag(p \mid Sx \mid^{p-2})$

Setting the gradient to zero, one gets,

$$\left(I + \frac{\lambda}{2a} S^T DS\right)x = b \tag{4.7}$$

Using the following matrix inversion lemma,

$$\left(I + \frac{\lambda}{2a} S^T DS\right)^{-1} = I - S^T\left(\frac{2a}{\lambda} D^{-1} + S^T S\right)^{-1} S \tag{4.8}$$

we arrive at the following identity,

$$x = b - S^T\left(\frac{2a}{\lambda} D^{-1} + S^T S\right)^{-1} Sb \tag{4.9}$$

or equivalently,

$$z = \left(\frac{2a}{\lambda} D^{-1} + S^T S\right)^{-1} Sb, x = b - S^T z \tag{4.10}$$

Solving z requires solving the following,

$$\left(\frac{2a}{\lambda} D^{-1} + S^T S\right)z = Sb \tag{4.11}$$

Adding cz to both sides of (4.11) and subtracting $S^T Sz$ gives,

$$z = \left(\frac{2a}{\lambda} D^{-1} + cI\right)^{-1} (cz_{k-1} + S(b - S^T z_{k-1})) \tag{4.12}$$

where c is the maximum eigenvalue of $S^T S$.

This derivation gives a three-step iterative solution to the unconstrained analysis prior optimization problem:

Initialize: $x_0 = 0, z_0 = 0$
Repeat until convergence
Step 1: $b = x_{k-1} + \frac{1}{a} A^T(y - Ax_{k-1})$
Step 2: $z_k = \left(\frac{2a}{\lambda} D^{-1} + cI\right)^{-1} (cz_{k-1} + A(b - A^T z_{k-1}))$, where $D = diag(p\,|Sb|^{p-2})$
Step 3: $x_k = b - S^T z_k$
End

This is the solution for the unconstrained problem. Theoretically, the constrained forms and the unconstrained forms are equivalent for correct choice of λ and ε. However, for all practical problems, it is not possible to find the relation between the two explicitly. Owing to the smoothness of the Pareto curve, solving the unconstrained problems for a decreasing sequence of λ is guaranteed to reach the solution desired by the constrained form. Based on this idea, one can use a cooling algorithm to solve the said problem by iteratively solving (4.5).

1. Choose a high value of λ initially.
2. Solve the unconstrained optimization (4.5) for the given value of λ.
3. Decrease the value of λ and go back to Step 1.
4. Continue Steps 2 and 3, until the $\|y - Ax\|_2^2 \le \varepsilon$.

Split Bregman

The MM technique was in vogue in the early days of compressed sensing. The problem with this approach was its slower convergence rate (empirically). Since the advent of variable splitting Bregman techniques, the MM algorithm has hardly been used.

We intend to solve the unconstrained problem (we know how the constrained form can be solved starting with the unconstrained one), that is,

$$\min_x \|y - Ax\|_2^2 + \lambda \|Sx\|_1 \tag{4.13}$$

One can substitute, $z = Sx$. The Lagrangian would be,

$$\min_{x,z} \|y - Ax\|_2^2 + \lambda \|z\|_1 \text{ such that } z = Sx \tag{4.14}$$

Imposing strict equality between the proxy and the variable in every step is an overkill. In practice, we only need to enforce strict equality. This leads to the concept of augmented Lagrangian, where the equality constraint is relaxed. This is given by,

$$\min_{x,z} \|y - Ax\|_2^2 + \lambda \|z\|_1 + \mu \| z - Sx\|_2^2 \tag{4.15}$$

Here, the parameter μ controls the strictness of equality. For small values, the equality is relaxed, and for large values, it is enforced. Ideally, one would have to solve (4.15) starting from small values of μ and eventually increase its value progressively. Such an approach is heuristic. A better way is to introduce a Bregman relaxation variable (b).

$$\min_{x,z} \|y - Ax\|_2^2 + \lambda \|z\|_1 + \mu \| z - Sx - b\|_2^2 \tag{4.16}$$

The variable is automatically updated in every iteration and hence controls the degree of strictness of the equality constraint. One can keep the parameter μ fixed for such Split Bregman formulations.

For solving (4.16), one typically follows the alternating direction method of multipliers (ADMM). The basic idea is to update each variable separately.

$$\min_{x} \|y - Ax\|_2^2 + \mu \| z - Sx - b\|_2^2 \tag{4.17}$$

$$\min_{z} \lambda \|z\|_1 + \mu \| z - Sx - b\|_2^2 \tag{4.18}$$

The update for x is a simple linear least squares problem, and the update for z is a l_1-minimization problem; we already know that (4.18) can be solved by one step of soft thresholding. Therefore, both (4.17) and (4.18) have closed-form solutions.

The final step in each iteration is to update the Bregman variable.

$$b \leftarrow z - Sx - b \tag{4.19}$$

The update of the Bregman variable controls the difference between the proxy and the variables. This allows the parameter μ to remain fixed.

Greedy Analysis Pursuit

Let us recall the strategy for solving synthesis prior problems greedily. The basic idea in algorithms such as orthogonal matching pursuit (OMP) is that the support of the sparse signal is estimated iteratively, from which the values at those non-zero positions are obtained.

In greedy analysis prior (GAP) algorithm, the strategy is opposite. We start from a dense co-sparse solution and progressively remove positions that do not satisfy the constraints. This is known as identifying the co-support, that is, position of zero values in the co-sparse signal.

GAP Algorithm

Initialize: $x_0 = \min_x \|Sx\|_2^2$ subject to $\|y - Ax\|_2^2$

Until convergence$\left(\|y - Ax\|_2^2 \geq \varepsilon\right)$

$x_k = \min_x \|S_{\Omega_k} x\|_2^2$ subject to $\|y - Ax\|_2^2$, Ω_k: support at kth iteration

$\Omega_{k+1} \leftarrow \Omega_k \setminus \left\{\arg\max\left\{|\omega^j x_k| : j \in \Omega_k\right\}\right\}, \omega^j$: jth row of S

Let us study each step of the algorithm. The initialization is straightforward. It is simply the minimum energy solution. In every iteration, we estimate the co-sparse signal x_k, given the current support. This is the minimum energy solution (at the current support) having a closed-form solution. After this step, one needs to update the support. Assuming that x^k is a good estimate of the actual signal x, it is reasonable to update the co-support (zero values in Sx^k) by small values of Sx^k. This is represented in the last step.

For the noiseless case, one needs to solve $\min_x \|Sx\|_2^2$ subject to $y = Ax$ in every iteration. The closed form for the same is derived here.

$$\min_x \|Sx\|_2^2 \text{ subject to } y = Ax$$

Lagrangian

$$L(x, \lambda) = \|Sx\|_2^2 + \lambda^T (y - Ax)$$

$$\nabla_x L(x,\lambda) = 2S^T Sx - A^T \lambda = 0 \text{ ... (1)}$$

$$\nabla_y L(x,\lambda) = y - Ax = 0 \text{ .... (2)}$$

From (1)

$$x = \frac{1}{2}\left(S^T S\right)^{-1} A^T \lambda \text{ .... (3)}$$

Putting the value of x in (2) for solving λ

$$y - \frac{1}{2}A\left(S^T S\right)^{-1} A^T \lambda = 0$$

$$\Rightarrow \lambda = 2\left(A\left(S^T S\right)^{-1} A^T\right)^{-1} y$$

Substitution λ in (3)

$$x = \left(S^T S\right)^{-1} A^T \left(A\left(S^T S\right)^{-1} A^T\right)^{-1} y$$

For the noisy case, one needs to solve $\min_x \|Sx\|_2^2$ subject to $\|y - Ax\|_2^2 \le \varepsilon$. This is difficult to solve. Therefore, in practice, one solves the unconstrained form of $\min_x \|y - Ax\|_2^2 + \lambda \|Sx\|_2^2$. This is the form of standard Tikhonov regularization.

$$\min_x \|y - Ax\|_2^2 + \lambda \|Sx\|_2^2$$

Differentiating

$$-A^T(y - Ax) + \lambda S^T Sx = 0$$

$$\Rightarrow \left(A^T A + \lambda S^T S\right)x = A^T y$$

$$\therefore x = \left(A^T A + \lambda S^T S\right)^{-1} A^T y$$

In practice, the noisy version is easier to solve. It can be solved efficiently by using conjugate gradient; it does not require explicit inverses such as the noise-free version. Hence, even if the system is noise free, the latter is solved by assuming a small value of λ.

Suggested Reading

I. W. Selesnick and M. A. Figueiredo, Signal restoration with overcomplete wavelet transforms: Comparison of analysis and synthesis priors. In *Wavelets XIII* (Vol. 7446, p. 74460D). International Society for Optics and Photonics, Washington, DC, 2009.

P. L. Combettes and J. C. Pesquet, Proximal splitting methods in signal processing. In *Fixed-Point Algorithms for Inverse Problems in Science and Engineering*, Springer, New York, pp. 185–212, 2011.

E. Ghadimi, A. Teixeira, I. Shames and M. Johansson, Optimal parameter selection for the alternating direction method of multipliers (ADMM): Quadratic problems. *IEEE Transactions on Automatic Control*, 60 (3), 644–658, 2015.

S. Nam, M. E. Davies, M. Elad and R. Gribonval, December. Recovery of cosparse signals with greedy analysis pursuit in the presence of noise. In *Computational Advances in Multi-Sensor Adaptive Processing (CAMSAP), 2011 4th IEEE International Workshop on*, IEEE, pp. 361–364, 2011.

Appendix: MATLAB® Codes

Majorization Minimization

```
function beta = LpAnalysis(H, Ht, A, At, y, p, normfac, err,
insweep, decfac, tol)

% Algorithm for solving problems of the form:
% min ||Ax||_p s.t. ||y-Hx||_2 < err

% Inputs
% A - Sparsifying Forward Transform
% At - Sparsifying Backward Transform
% H - Forward Measurement Operator
% Ht - Backward Measurement Operator
% y - collected data
% p - non-convex norm
% normfac - highest singular value of A (default 1)
% err - two norm mismatch (default 1e-6)
% insweep - number of sweeps for internal loop (default 50)
% decfac - lambda decrease factor for outside loop
(default 0.5)
% tol - tolerance for convergence (default 1e-4)

% copyright (c) Angshul Majumdar 2010

c = 1.1;

if nargin < 6
    normfac=1;
end
if nargin < 7
    p = 1;
end
if nargin < 8
    err = 1e-6;
end
if nargin < 9
    insweep = 50;
end
if nargin < 10
    decfac = 0.5;
end
```

```
if nargin < 11
    tol = 1e-4;
end

alpha = 1.1*normfac;
x_initial = Ht(y);
z_initial = zeros(length(A(x_initial)),1);

x = x_initial; z = z_initial;

lambdaInit = decfac*max(abs(Ht(y))); lambda = lambdaInit;

f_current = norm(y-H(x_initial)) + lambda*norm(A(x),p);

while lambda > lambdaInit*tol
    % debug lambda
    for ins = 1:insweep
        f_previous = f_current;

        b = x + (1/alpha)*(Ht(y-H(x)));
        z = (c*z + A(b-At(z)))./((2*alpha/
lambda)*(abs(A(x))).^(2-p)+c);
        x = b - At(z);

        f_current = norm(y-H(x)) + lambda*norm(A(x),p);

        if norm(f_current-f_previous)/norm(f_current +
f_previous)<tol
            break;
        end
    end
    if norm(y-H(x))<err
        break;
    end
    lambda = decfac*lambda;
end
beta = x;
```

Split Bregman Algorithm

```
function x =SB(y,A,S,b,lam,max_itr)
% Inputs
% y: measurement vector
% A: measurement operator (explicit matrix)
% S: sparsifying operator

%tuning is required for:
%b--------bregman variable
%lam--------value of lambda
%max_itr-------maximum iterations

% Output
% x: co-sparse signal

% initializations
x=randn(size(A,2),size(y,2)); z=randn(size(S,1),size(y,2));

cost=[];

for i=1:max_itr
    yt = [y ; (z-b)];At=[A;S];
    x  = At\yt;

    z  = soft(S*x+b, lam/2);

    b=(z-S*x-b);

    cost(i) =
norm((y-A*x),'fro')+norm((z-S*x-b),'fro')+lam*norm(z,1);
end


% Soft function

    function  z = soft(s,thld)
        z = sign(s).*max(0,abs(s)-thld);
    end
end
```

Greedy Analysis Pursuit

```
function [x] = GAPn(y, M, omega, eps)
% Solves problem of the form min ||omega*x||_0 subject
to y=Mx
% Input
% measurement y
% measurement matrix M
% sparsifying operator omega: pxd
% noise floor eps
%Output
% co-sparse signal x
p=size(omega,1);

st=1e-2;%5e-4 %%% can be varied
lambda=1; %%% can be varied
defac=0.7; %%% can be varied

i=0;
cosupport= 1:p ;
x= mldivide([M; sqrt(lambda)*omega(cosupport,:)], [y ; zeros(p,1)
]);

norm(y-M*x,'fro')
while(norm(y-M*x,'fro') >=eps)
    stopping_condition=true;
    while( stopping_condition)

    x= mldivide([M; sqrt(lambda)*omega(cosupport,:)] , [y ;
zeros (length (cosupport),1 ) ] ); %apply cooling technique
for cooling lambda

    [value, ind]= max( abs( omega(cosupport,:)*x ) );
    cosupport= cosupport( cosupport~=ind );
    i=i+1;

    stopping_condition= norm( omega(cosupport,:) *x,inf) > st
    end
lambda=defac*lambda
end

end
```

5

Group Sparsity

The problem of group sparsity was briefly introduced in the introductory chapter. We have studied about algorithms on sparse recovery in detail. This is the most generic problem. We only know that the solution is sparse. But what if we have more information about the solution? Can we have a better solution? These are topics of model-based compressed sensing.

For example, consider the problem of analyzing an image in wavelet domain; the wavelet coefficients are sparse, but they are not arbitrarily sparse. They have a tree structure, that is, if the wavelet coefficient at a higher level is non-zero, its corresponding coefficient at a lower-level is also non-zero. Thus, the sparse coefficients have a tree-like structure. This is one example of model-based compressed sensing.

Another such model that is more ubiquitous in signal processing and machine learning is group sparsity. Here, we know that there are some groups that are always non-zero simultaneously; that is, if the group is non-zero, all the coefficients within that group are non-zeroes. There is never a case where some of the coefficients within the group are zeroes and the others are non-zeroes. We studied one example of signal processing in the introductory chapter; we will learn more about its application in machine learning later. Here, our interest is in solving such problems.

Synthesis Prior

As usual, we will first study the group-sparse synthesis prior recovery problems, simply because they are more in vogue. At a later stage, we will discuss solution to analysis prior problems.

Solving the Equality-Constrained Problem

The equality-constrained group-sparsity recovery problem is posed as follows,

$$\min_{x} \|x\|_{2,1} \text{ subject to } y = Ax \tag{5.1}$$

The $l_{2,1}$-norm is defined as the sum of the l_2-norms of the groups. The groups need to be pre-defined.

We will use the Focal Underdetermined System Solver (FOCUSS) approach to solve this.

The Lagrangian for (5.1) is

$$L(x,\lambda) = \|x\|_{2,1} + \lambda^T(Ax - y) \tag{5.2}$$

Following the theory of Lagrangian, the stationary point for (5.2) needs to be obtained by solving

$$\nabla_x L(\hat{x},\hat{\lambda}) = 0 \tag{5.3}$$

$$\nabla_\lambda L(\hat{x},\hat{\lambda}) = 0 \tag{5.4}$$

Solving (5.3) requires the gradient of $E(x) = \|x\|_{2,1}$ with respect to x. In general, the gradient of $E(x)$ can be expressed as,

$$\nabla_x E(x) = \alpha(x)\Pi(x)x \tag{5.5}$$

where $\alpha(x)$ is a scalar, and $\Pi(x)$ is a diagonal matrix.

Solving (5.3) yields (5.6) and solving (5.4) yields (5.7),

$$\hat{x} = -\frac{1}{\alpha(\hat{x})}\Pi(\hat{x})^{-1}A^T\hat{\lambda} \tag{5.6}$$

$$\hat{\lambda} = -\alpha(x)(A\Pi(\hat{x})^{-1}A^T)^{-1}y \tag{5.7}$$

Substituting (5.7) into (5.6) yields,

$$\hat{x} = \Pi(\hat{x})^{-1}A^T(A\Pi(\hat{x})^{-1}A^T)^{-1}y \tag{5.8}$$

The expression (5.8) is an implicit function of (x). Therefore, it needs to be solved iteratively in the following manner:

$$x(k+1) = \Pi(x(t))^{-1}A^T(A\Pi(x(k))^{-1}A^T)^{-1}y \tag{5.9}$$

The diversity measure for the group-sparse optimization problem is $\|x\|_{2,1} = \left(\sum_{i=1}^{C} \|x_i\|_2\right)$, and the gradient is

$$\nabla_x = \left[\frac{\partial}{\partial x_{1,1}}, \ldots, \frac{\partial}{\partial x_{1,n_1}}, \ldots, \frac{\partial}{\partial x_{i,j}}, \ldots, \frac{\partial}{\partial x_{C,1}}, \ldots, \frac{\partial}{\partial x_{C,n_C}} \right]^T$$

where $\partial \|x\|_{2,1} / \partial x_{i,j} = \|x_i\|_2^{-1} |x_{i,j}|^{-1} x_{i,j}$

Comparing this with (5.5), we have,

$$\alpha(x) = 1$$
$$\Pi(x) = diag\left(\|x_i\|_2^{-1} |x_{i,j}|^{-1} \right) \tag{5.10}$$

In principle, we can apply (5.10) repeatedly to solve our problem. This is a sparse optimization problem, and therefore, the groups and the individual coefficients can be zeroes; inverting them would lead to numerical instability. To address this, we add a damping factor,

$$\Pi(x) = diag\left(\|x_i\|_2^{-1} |x_{i,j}|^{-1} + \delta \right) \tag{5.11}$$

where δ is the damping factor, whose value is reduced at each iteration.

Computing (5.8) at each step is computationally expensive, since it involves finding explicit matrix inverses. To avoid this expense, one can employ the conjugate gradient (CG) method, in the following steps:

$$R = diag\left[\left(\|x_i\|_2^{-1} |x_{i,j}|^{-1} + \delta \right)^{-1/2} \right] \tag{5.12}$$

$$\Phi = AR \tag{5.13}$$

$$z(k+1) = \min \|y - \Phi z\|_2 \text{, by CG method} \tag{5.14}$$

$$x(k+1) = Rz(k+1) \tag{5.15}$$

This solution precludes computing explicit inverses. Moreover, it does not require A to be specified explicitly; it can also be applied as fast operators.

Solving the Unconstrained Problem

We have already learnt how to solve the l_1-norm minimization problem. This is expressed as:

$$\min_x \|y - Ax\|_2^2 + \lambda \|x\|_1 \tag{5.16}$$

Here, we have seen how the $l_{2,1}$-norm can be handled. The unconstrained $l_{2,1}$-norm minimization problem is expressed as,

$$\min_x \|y - Ax\|_2^2 + \lambda \|x\|_{2,1} \tag{5.17}$$

We follow the same procedure as before, that is majorization minimization (MM). In every iteration, we majorize $\|y - Ax\|_2^2$; this leads to the following minimization problem,

$$x_k = \min_x \|b - x\|_2^2 + \lambda \|x\|_{2,1} \tag{5.18}$$

where $b = x_{k-1} + 1/\alpha\, A^T(y - Ax)$ and α is the maximum Eigenvalue of A^TA.

The gradient of (5.18) is expressed as $2(b - x) + \lambda \|x_i\|_2^{-1} x_{i,j}$.

This can be expressed in our familiar form,

$$2b - 2x + \frac{\lambda}{\alpha}\Lambda signum(x), \text{ where } \Lambda = diag\left(\|x_i\|_2^{-1} x_{i,j}\right) \tag{5.19}$$

This allows us to update x via modified iterative soft thresholding.

$$x_k = signum(b)\max\left(0, |b| - \frac{\lambda}{2a}\Lambda\right) \tag{5.20}$$

Smoothed $l_{2,0}$-Minimization

Just as l_1-minimization is an approximation of the ideal l_0-minimization, the $l_{2,1}$-norm is a convex approximation of the $l_{2,0}$-norm in case of group sparsity. The $l_{2,0}$-norm is defined as the number of groups having non-zero l_2-norms. As you expect, this is a non-deterministic polynomial time (NP) hard problem. However, one can have an approximate solution.

The smooth l_0-norm minimization is proposed as a solution to the NP hard sparse estimation problem. It replaced the NP hard (spiky) l_0-norm by a parameterized smooth function, whose smoothness could be varied. The smooth function allowed for the usage of gradient-based optimization techniques. The solution was achieved iteratively by gradually reducing the smoothness of the function. This is a typical homotopy approach. We can adopt a similar idea for solving the $l_{2,0}$-norm minimization problem (5.21),

$$\min_x \|x\|_{2,0} \text{ such that } y = Ax \tag{5.21}$$

$$\text{We define: } \gamma(x_i) = \begin{cases} 1 \text{ when } \|x_i\|_2 > 0 \\ 0 \text{ when } \|x_i\|_2 = 0 \end{cases}$$

Based on the above definition,

$$\|x\|_{2,0} = \sum_{i=1}^{C} \gamma(x_i) \tag{5.22}$$

In (5.21), we replace the non-smooth (spiky) function $\gamma(x)$ by a smooth zero-mean Gaussian function whose smoothness can be varied. We define,

$$f_\sigma(x_i) = e^{(-\|x_i\|_2^2/2\sigma^2)} \tag{5.23}$$

The function is smooth when σ is large and becomes spiky when its value reduces. In the limit that the σ is zero, the above function has the following property,

$$\lim_{\sigma\to 0} f_\sigma(x_i) = \begin{cases} 1 \text{ when } \|x_i\|_2 = 0 \\ 0 \text{ when } \|x_i\|_2 > 0 \end{cases} \tag{5.24}$$

Therefore, $\lim_{\sigma\to 0} f_\sigma(x_i) = 1 - \gamma(x_i)$. This allows for approximating the $l_{2,0}$-norm by,

$$\lim_{\sigma\to 0} F_\sigma(x) = \sum_{i=1}^{C} f_\sigma(x_i) = \sum_{i=1}^{C} 1 - \gamma(x_i) = N - \|x\|_{2,0} \tag{5.25}$$

Therefore, the $l_{2,0}$-norm minimization problem can be recast as

$$\max \lim_{\sigma \to 0} F_\sigma(x) \text{ subject to } y = Ax \tag{5.26}$$

Since the objective function is smooth, it is easy to solve (5.26) by gradient-based methods. The main idea behind the algorithm is that, at each iteration, (5.26) is solved for a particular value of σ; in the following iteration, the value of σ is decreased and (5.26) is solved again. This continues till the solution has converged.

SMOOTHED $l_{2,0}$-NORM MINIMIZATION

Initialization: Obtain the initial solution $\hat{x}^{(0)} = \min_x \|y - Ax\|_2^2$.
At Iteration k: Continue the following steps till solution is reached (i.e., till σ is greater than a specified value)
Choose $\sigma = c \cdot \max(\|x_i\|_2)$, where c is a constant greater than 4.
Maximize (5.26) for the current value of σ. Steepest ascent method is used to achieve this.
Initialize, $s = x^{(k-1)}$.
Let $\Delta s = [s_1 \cdot e^{-\|s_1\|_2^2/2\sigma^2}, \ldots, s_n \cdot e^{-\|s_n\|_2^2/2\sigma^2}]^T$
Update: $s \leftarrow s - \mu \Delta s$, where μ is a small constant.
Project the solution back to the feasible set by $s \leftarrow s - A^T(AA^T)^{-1}(As - x^{(k-1)})$. To avoid calculating explicit inverses, the projection is broken down into the following three steps:
Compute: $t_1 = As - x^{(k-1)}$.
Solve: $\hat{t}_2 = \min \|t_1 - At_2\|_2$.
Project: $s \leftarrow s - \hat{t}_2$.
Set $x^{(k)} = s$, and return to step 1 until convergence.

Analysis Prior

The analysis prior formulation assumes that the signal is not sparse but has a sparse representation in another basis (or transform). Unlike the synthesis prior formulation, it recovers the signal itself rather than the transform coefficients. We are interested in solving the following optimization problem,

$$\min_x \|y - Ax\|_2^2 + \lambda \|Hx\|_{2,1} \tag{5.27}$$

As before, we use the MM approach. The function to be minimized is,

$$J(x) = \|y - Ax\|_2^2 + \lambda \|Hx\|_{2,1} \tag{5.28}$$

There is no closed-form solution to $J(x)$; it must be solved iteratively. At each iteration, we choose the majorizer to be,

$$G_k(x) = \|y - Ax\|_2^2 + (x - x_k)^t (aI - A^T A)(x - x_k) + \lambda \|Hx\|_{2,1} \tag{5.29}$$

$G_k(x)$ satisfies the condition for the MM algorithm when a is greater than the maximum eigenvalue of $A^T A$.

Now, $G_k(x)$ can be alternately expressed as follows,

$$G_k(x) = a \left\| x_k + \frac{1}{a} A^T (y - Ax) - x \right\|_2^2 + \lambda \|Hx\|_{2,1} + K \tag{5.30}$$

where K is a term independent of x.

Minimizing (5.30) is the same as minimizing the following,

$$\|b - x\|_2^2 + \frac{\lambda}{a} \|Hx\|_{2,1} \tag{5.31}$$

where $b = x_k + \frac{1}{a} A^T (y - Ax_k)$.

Now, the problem is to minimize (5.31). To do so, we take the derivative of the function,

$$\nabla G_k'(x) = 2x - 2b + (\lambda / a) H^T D H x \tag{5.32}$$

where $D = Diag(\|Hx^{(i)}\|_2^{-1})$. The "." denotes element-wise product.

Setting the gradient to zero, one gets,

$$\left(I + \frac{\lambda}{2a} H^T D H \right) x = b \tag{5.33}$$

It is not possible to solve (5.33) directly. This is because, in all practical situations, the sparsifying transform H is available as a fast operator, not as an explicit matrix. It is possible to form the corresponding matrices, but for real signals, such matrices will be so large (leading to huge memory and computational requirements) that it would be impossible to solve (5.33) on a PC. To overcome this problem, the matrix inversion lemma is used to simplify (5.33).

$$\left(I+\frac{\lambda}{2a}H^TDH\right)^{-1}=I-H^T\left(\frac{2a}{\lambda}D^{-1}+H^TH\right)^{-1}H$$

From (5.33), we have, using the above identity,

$$x=b-H^T\left(\frac{2a}{\lambda}D^{-1}+H^TH\right)^{-1}Hb$$

Equivalently,

$$z=\left(\frac{2a}{\lambda}D^{-1}+H^TH\right)^{-1}Hb,\ x=b-H^Tz \tag{5.34}$$

Solving z requires the solution of the following,

$$\left(\frac{2a}{\lambda}D^{-1}+HH^T\right)z=Hb$$

Adding cz to both the right and left sides and subtracting HH^Tz gives,

$$\left(\frac{2a}{\lambda}D^{-1}+cI\right)z=Hb+(cI-HH^T)z$$

This suggests the following update,

$$z_{k+1}=\left(\frac{2a}{\lambda}D^{-1}+cI\right)^{-1}(cz_k+H(b-H^Tz_k)) \tag{5.35}$$

$$x_{k+1}=b-H^Tz_{k+1} \tag{5.36}$$

where $b = x_k + \frac{1}{a} P^T (y - P x_k)$.

The value of c should be chosen such that it is larger than the maximum eigenvalue of $H^T H$.

Greedy Algorithms

The smoothed $l_{2,0}$-norm minimization problem solves the ideal problem approximately. Here, we will learn techniques to solve the problem greedily. We are interested in solving

$$\min_x \|x\|_{2,0} \text{ such that } y = Ax$$

where $x = [\underbrace{x_{1,1},\ldots,x_{1,n_1}}_{x_1},\underbrace{x_{2,1},\ldots,x_{2,n_2}}_{x_2},\ldots\underbrace{x_{k,1},\ldots,x_{k,n_k}}_{x_k}]^T$

We have studied greedy algorithms in a previous chapter. In that chapter, we explained in detail the ideas behind the same. Here, we will only discuss how these can be changed to solve group-sparse recovery problems. The basic idea remains the same. In the prior method, we recovered unstructured sparse coefficients; here, we will be selecting blocks of coefficients.

First, we modify the orthogonal matching pursuit (OMP) algorithm; it will be called the block OMP (BOMP).

BOMP ALGORITHM

Given – Matrix A, vector y, and group labels of each column in A
Required – Estimate a block sparse vector x
Initialize – iteration $t = 0$; residual $r_0 = y$; support $\Omega_0 = []$
Repeat until a stopping criterion is met

1. Compute : $c_t = A^T r_{t-1}$
2. Select support : $l_t = \{j : \text{group}(\max(c_t^j))\}$
3. Update support : $\Omega_t = \Omega_{t-1} \cup l_t$
4. Estimate values at support : $x_t = \arg\min_x \|y - A_{\Omega_t} x\|_2^2$
5. Update residual : $r_t = y - Ax_t$

The BOMP algorithm is similar to the OMP algorithm, except for the support selection step. In OMP, we are to select only the position of one non-zero value; here, we want to select one non-zero group. There are two ways to do that. In BOMP, we assume that the group corresponding to the highest non-zero value is one of the active/non-zero groups; therefore, we select the entire group. This is the only difference from the OMP.

One can have an alternate selection strategy. Instead of looking at only one value, we can look at the values in the entire group. We can select the group having the maximum l_2-norm. This is more in line with the definition of the $l_{2,0}$-norm. With this selection criteria, we have the group OMP (GOMP).

GOMP ALGORITHM

Given – Matrix A, vector y, and group labels of each column in A
Required – Estimate a block sparse vector x
Initialize – iteration $t = 0$; residual $r_0 = y$; support $\Omega_0 = []$
Repeat until a stopping criterion is met

1. Compute: $c_t = A^T r_{t-1}$
2. Select support: $l_t = \{j : \text{group}(\max \|c^j\|_2)\}$
3. Update support: $\Omega_t = \Omega_{t-1} \cup l_t$
4. Estimate values at support: $x_t = \arg\min_x \|y - A_{\Omega_t} x\|_2^2$
5. Update residual: $r_t = y - Ax_t$

The OMP-based algorithms require the solution of a least squares problem (Step 4). Depending upon the size of the problems, this may be expensive. Now, we will discuss some algorithms that reduce the computational cost of the algorithm by either skipping that step (matching pursuit) or approximately solving that step (gradient pursuit).

As before, there can be two versions of matching pursuit. Depending on how we are selecting the group, there can be a block version (selection based on the highest value of individual coefficient) or a group version (selection based on the highest l_2-norm of group). Here, we show only one; the reader can easily derive/implement the other.

BLOCK MATCHING PURSUIT ALGORITHM

Given – Matrix A, vector y, and group labels of each column in A
Required – Estimate a block sparse vector x
Initialize – iteration $t = 0$; residual $r_0 = y$
Repeat until a stopping criterion is met

1. Compute : $c_t = A^T r_{t-1}$
2. Select support : $l_t = \{j : \text{group}(\max(c_t^j))\}$
3. Update variable : $x_t(l_t) = x_{t-1}(l_t) + c_t(l_t)$
4. Update residual : $r_t = r_{t-1} - A_{l_t} x_t(l_t)$

The first step is similar to the OMP-based algorithms. It calculates the correlations between the current residual and the columns of A^T. In the next step, it selects the group indices having the highest correlation according to the "Block" selection criterion. In Step 3, it only updates the sparse estimate at the group indices selected in the previous step. Step 4 updates the residual by subtracting the product of the matrix A and the sparse estimate x indexed in Step 2 from the previous residual. As before, we can also have a group matching pursuit (GMP), in which the indices are selected in Step 2 according to the "Group" selected method, that is, selecting all indices of the group having the highest l_2-norm.

Although the MP-based algorithms are very fast, they are relatively inaccurate compared with those based on OMP. There is always a trade-off between computational expense and accuracy. There is a middle path between the two—gradient pursuit algorithms. These are based on approximately solving the least squares problem in OMP to keep the computational cost low (but at the same time does not compromise much on the accuracy). However, these are not predominant techniques and will not be discussed any further.

There is a better way to reduce the overall complexity of the algorithm. In OMP, only a single non-zero support is detected. If we are greedier and choose more than one, we can terminate the iterations faster. This is achieved by selecting multiple indices in each iteration instead of selecting only one in the sparse recovery problem. One can adopt such super-greedy approaches for the block/group-sparse selection criteria as well. Here, we look at two algorithms: stagewise BOMP (StBOMP) and stagewise weak group OMP (SwGOMP).

StBOMP ALGORITHM

Given – Matrix A, vector y, and group labels of each column in A
Required – Estimate a block sparse vector x
Initialize – iteration $t = 0$; residual $r_0 = y$; support $\Omega_0 = []$
Repeat until a stopping criterion is met

1. Compute : $c_t = A^T r_{t-1}$
2. Select support : $l_t = \{j : \text{group}(c_t^j > \tau)\}$
3. Update support : $\Omega_t = \Omega_{t-1} \cup l_t$
4. Estimate values at support : $x_t = \arg\min_x \|y - A_{\Omega_t} x\|_2^2$
5. Update residual : $r_t = r_{t-1} - A_{l_t} x_t(l_t)$

where $\tau = 2\|r_t\|_2 / n^{\frac{1}{2}}$, n being the length of y.

StBOMP is a greedier algorithm than BOMP; at the Selection Step 2, it greedily selects more than one group at a time. It selects all the groups that have correlations greater than a threshold. Owing to greedier selection, it terminates faster than the regular BOMP.

Our last algorithm is the SwGOMP. It is based on modifying the stagewise weak OMP (SwOMP). In SwOMP, instead of choosing only the highest value, one chooses all indices that are within a range, say $\lambda = 70\%$ of the highest value. One can adopt the same criteria for group-sparse recovery.

SwGOMP ALGORITHM

Given – Matrix A, vector y, and group labels of each column in A.
Required – Estimate a block sparse vector x
Initialize – iteration $t = 0$; residual $r_0 = y$; support $\Omega_0 = []$
Repeat until a stopping criterion is met

1. Compute : $c_t = A^T r_{t-1}$
2. Select support : $l_t = \{j : \text{group}(c^j > \lambda \max \|c^j\|_2)\}$
3. Update support : $\Omega_t = \Omega_{t-1} \cup l_t$
4. Estimate values at support : $x_t = \arg\min_x \|y - A_{\Omega_t} x\|_2^2$
5. Update residual : $r_t = r_{t-1} - A_{l_t} x_t(l_t)$

where $\tau = 2\|r_t\|_2 / n^{\frac{1}{2}}$, n being the length of y.

There can be many variations to greedy group-sparse recovery techniques. We have discussed some of the popular ones. Once the reader familiarizes himself/herself with these techniques, they will be in a position to modify any greedy sparse recovery algorithm to solve the group-sparsity problem.

Suggested Reading

D. L. Donoho, Y. Tsaig, I. Drori and J. L. Starck, Sparse solution of underdetermined systems of linear equations by stagewise orthogonal matching pursuit, *IEEE Transactions on Information Theory*, 58 (2), 1094–1121, 2012.

Y. C. Eldar, P. Kuppinger and H. Bolcskei, Block-sparse signals: Uncertainty relations and efficient recovery, *IEEE Transactions on Signal Processing*, 58 (6), 3042–3054, 2010.

A. Majumdar and R. K. Ward, Fast group sparse classification, *Canadian Journal of Electrical and Computer Engineering*, 34 (4), 136–144, 2009.

Appendix: MATLAB® Codes

Reweighted Least Square/FOCUSS Implementation

```
function x = rwlsmpq(A,y,group,m,p,q)

% Solution to the non-convex optimization problem
min||x||_m,p subject to
% ||y - Ax||_q < eps
% This algorithm is based upon the Reweighted Least-
squares algorithm
%
% Copyright (c) Angshul Majumdar 2009

% Input
% A = N X d dimensional measurement matrix
% y = N dimensional observation vector
% m - inner norm for group (default 2)
% p - norm for sparsity of group (default 1)
% q - norm for model fitting (default 2)

% Output
% x = estimated sparse signal

if nargin < 4
    m = 2; p = 1; q = 2;
end
if nargin < 5
    p = 1; q = 2;
end
if nargin < 6
    q = 2;
end

explicitA = ~(ischar(A) || isa(A, 'function_handle'));
if (explicitA)
    AOp = opMatrix(A); clear A
else
    AOp = A;
end

% Set LSQR parameters
damp   = 0;
atol   = 1.0e-6;
```

```
btol   = 1.0e-6;
conlim = 1.0e+10;
itnlim = length(y);
show   = 0;
OptTol = 1e-6;

MaxIter = 500;
epsilon = 1;
NGroup = max(group);
for i = 1:NGroup
    GInd{i} = find(group == i);
end
% u_0 is the L_2 solution which would be exact if m = n,
% but in Compressed expectations are that m is less than n

[u1,u_0] = lsqr(@lsqrAOp,y,OptTol,20);
u_old = u_0; % use regularized estimate
j=0;
while (epsilon > 1e-5) && (j < MaxIter)
    j = j + 1;
    for i = 1:NGroup
        tw1(GInd{i}) = norm(u_old(GInd{i})).^(p-m);
    end
    tw2 = abs(u_old).^(m-2);
    wm = tw1'.*tw2 + epsilon;
    vm = 1./sqrt(wm);
    Wm = opDiag(vm);
    wd = (2/q)*(((AOp(u_old,1)-y).^(2) + epsilon).^(q/2-1));
    % data fidelity term
    vd = 1./sqrt(wd);
    Wd = opDiag(vd);

    MOp = opFoG(Wd,AOp,Wm);
    ybar = Wd(y,1);
    [t1,t] = lsqr(@lsqrMOp,ybar,OptTol,20); % use regularized
    estimate
    u_new = Wm(t,1);
    if lt(norm(u_new - u_old,2),epsilon^(1/2)/100)
        epsilon = epsilon /10;
    end
    u_old = u_new;
end
x = u_new;


%%%%%%%%%%%%%%%%%%%%%%%%%%%%%%%%%%%%%%%%%%%%%%%%%%%%%
```

```
function y = lsqrAOp(x,transpose)
        switch transpose
            case 'transp'
                y = AOp(x,2);
            case 'notransp'
                y = AOp(x,1);
        end
    end
    function y = lsqrMOp(x,transpose)
        switch transpose
            case 'transp'
                y = MOp(x,2);
            case 'notransp'
                y = MOp(x,1);
        end
    end
end
```

Smoothed $l_{2,0}$-minimization

```
function s=SL20(A, x, group, sigma_min)

% Solution to the optimization problem min||s||_2,0
subject to x = As
% This is an NP hard problem. It is solved iteratively by
smoothing the
% mixed ||.||_2,0 norm
% This algorithm is based upon the SL0 algorithm from the
following webpage
%
% Web-page:
% ------------------
%    http://ee.sharif.ir/~SLzero
%
% % Copyright (c) Angshul Majumdar 2009

% Input
% A = N X d dimensional measurment matrix
% y = N dimensional observation vector
% group = labels

% Output
% s = estimated sparse signal

MaxIter = 2500; j = 0;
    sigma_decrease_factor = 0.5;
```

```
    A_pinv = pinv(A);
    mu_0 = 2;
    L = 3;
    ShowProgress = logical(0);

NoOfGroups = length(unique(group));
for i = 1:NoOfGroups
    GroupInd{i} = find(group == i);
end

% Initialization
%s = A\x;
s = A_pinv*x;
sigma = 2*max(abs(s));

% Main Loop
while (sigma>sigma_min) && (j < MaxIter) % && (norm(x-
A*s) > err)
    j = j + 1;
    for i=1:L
        delta = OurDelta(s,sigma,GroupInd);
        s = s - mu_0*delta';
        s = s - A_pinv*(A*s-x);    % Projection
    end

    if ShowProgress
        fprintf('     sigma=%f, SNR=%f\n',sigma,estimate_
        SNR(s,true_s))
    end

    sigma = sigma * sigma_decrease_factor;
end


%%%%%%%%%%%%%%%%%%%%%%%%%%%%%%%%%%%%%%%%%%%%%%%%%%%%%%%%%%%%
function delta=OurDelta(s,sigma,GroupInd)
for i = 1:length(GroupInd)
    delta(GroupInd{i}) = s(GroupInd{i}).*exp(-norm
    (s(GroupInd{i}))^2/(2*sigma^2));
end

%%%%%%%%%%%%%%%%%%%%%%%%%%%%%%%%%%%%%%%%%%%%%%%%%%%%%%%%%%%%
function SNR=estimate_SNR(estim_s,true_s)

err = true_s - estim_s;
SNR = 10*log10(sum(true_s.^2)/sum(err.^2));
```

BOMP

```
function [s, residual] = BOMP(A, y, group, err)

% Block Orthogonal Matching Pursuit - selection of group
based on highest
% correlation of each group [1]

% Input
% A = N X d dimensional measurement matrix
% y = N dimensional observation vector
% group = labels

% Output
% s = estimated sparse signal
% r = residual

% [1] Y. C. Eldar and H. Bolcskei, "Block Sparsity:
Uncertainty  Relations
% and Efficient Recovery," to appear in ICASSP 2009

if nargin < 5
     err = 1e-5;
end

c = max(group);
s = zeros(size(A,2),1);
r(:,1) = y; L = []; Psi = [];
for j = 1:c
    g{j} = find(group == j);
end
i = 2;

while (i < c) && (norm(r(:,end))>err)
    l = A'*r(:,i-1);
    [B, IX] = sort(abs(l),'descend');
    I = g{group(IX(1))};
    L = [L' I']';
    Psi = A(:,L);
    x = Psi\y;
    yApprox = Psi*x;
    r(:,i) = y - yApprox;
    i = i + 1;
end

s(L) = x;
residual = r(:,end);
```

GOMP

```
function [s, residual] = GOMP(A, y, group, err)

% Group orthogonal Matching Pursuit - selection of group
based on highest
% average correlation of each group

% Input
% A = N X d dimensional measurment matrix
% y = N dimensional observation vector
% group = labels

% Output
% s = estimated sparse signal
% residual = residual

% Copyright (c) Angshul Majumdar 2009

if nargin < 5
    err = 1e-5;
end

c = max(group);
s = zeros(size(A,2),1); t = 2.5;
r(:,1) = y; L = []; Psi = [];
i = 2;
for j = 1:c
    g{j} = find(group == j);
end

while (i < c) && (norm(r(:,end))>err)
    l = A'*r(:,i-1);
    for j = 1:c
        lg(j) = mean(abs(l(g{j})));
    end
    [B, IX] = sort(lg, 'descend');
    L = [L' g{IX(1)}']';
    Psi = A(:,L);
    x = Psi\y;
    r(:,i) = y - Psi*x;
    i = i+1;
end

s(L) = x;
residual = r(:,end);
```

BMP

```
function [s, residual] = BMP(A, y, group, err)

% Block Matching Pursuit

% Input
% A = N X d dimensional measurment matrix
% y = N dimensional observation vector
% group = labels
% m = sparsity of the underlying signal

% Output
% s = estimated sparse signal
% r = residual

% Copyright (c) Angshul Majumdar 2009

if nargin < 5
    err = 1e-5;
end

c = max(group);
s = zeros(size(A,2),1);
r(:,1) = y; L = []; Psi = [];
i = 2;
for j = 1:c
    g{j} = find(group == j);
end
while (i < c) && (norm(r(:,end))>err)
    l = A'*r(:,i-1);
    [B, IX] = sort(abs(l), 'descend');
    j = 1;
    while isempty(find(g{j}==IX(1))) %#ok<EFIND>
        j = j+1;
    end
    s(g{j}) = s(g{j}) + l(g{j});
    mask = zeros(size(A,2),1);
    mask(g{j}) = l(g{j});
    r(:,i) = r(:,i-1) - A*mask;
    i = i+1;
end
residual = r(:,end);
```

GMP

```
function [s, residual] = GMP(A, y, group, err)

% Group Matching Pursuit

% Input
% A = N X d dimensional measurment matrix
% y = N dimensional observation vector
% group = labels
% m = sparsity of the underlying signal

% Output
% s = estimated sparse signal
% r = residual

% Copyright (c) Angshul Majumdar 2009

if nargin < 5
     err = 1e-5;
end

c = max(group);
s = zeros(size(A,2),1);
r(:,1) = y; L = []; Psi = [];
i = 2;
for j = 1:c
    g{j} = find(group == j);
end
while (i < c) && (norm(r(:,end))>err)
    l = A'*r(:,i-1);
    for j = 1:c
        lg(j) = mean(abs(l(g{j})));
    end
    [B, IX] = sort(lg, 'descend');
    L = [L' g{IX(1)}']';
    s(g{IX(1)}) = s(g{IX(1)}) + l(g{IX(1)});
    mask = zeros(size(A,2),1);
    mask(g{IX(1)}) = l(g{IX(1)});
    r(:,i) = r(:,i-1) - A*mask;
    i = i+1;
end
residual = r(:,end);
```

StGOMP

```
function [s, residual] = StGOMP(A, y, group, steps, err)

% Stagewise Group Orthogonal Matching Pursuit - Combining
Ideas from [1]
% and [2]

% Input
% A = N X d dimensional measurment matrix
% y = N dimensional observation vector
% group = labels
% steps = sparsity of the underlying signal

% Output
% s = estimated sparse signal
% r = residual

% Copyright (c) Angshul Majumdar 2009

% [1] Y. C. Eldar and H. Bolcskei, "Block Sparsity:
Uncertainty Relations
% and Efficient Recovery," to appear in ICASSP 2009
% [2] D.L. Donoho, Y. Tsaig, I. Drori, J.-L. Starck,
"Sparse solution of
% underdetermined linear equations by stagewise orthogonal
matching pursuit"
% preprint http://www-stat.stanford.edu/~idrori/StOMP.pdf

 if nargin < 5
     err = 1e-5;
 end
 if nargin < 4
     err = 1e-5;
     steps = 5;
 end

c = max(group);
s = zeros(size(A,2),1); t = 0.5;
r(:,1) = y; L = []; Psi = [];
i = 2;
for j = 1:c
    g{j} = find(group == j);
end
```

```
while (i < steps) && (norm(r(:,end))>err)
    l = sqrt(length(y)).*A'*r(:,i-1)./norm(r(:,i-1));
    thr = fdrthresh(l, t);
    l = find(abs(l)>thr);
    gr = unique(group(l));

    for k = 1:length(gr)
        L = [L' g{gr(k)}']';
    end
    Psi = A(:,L);
    x = Psi\y;
    r(:,i) = y - Psi*x;
    i = i+1;
end

s(L) = x;
residual = r(:,end);
```

Related Code

```
function thresh = fdrthresh(z,q)

% Copyright (c) 2006. Iddo Drori

% Part of SparseLab Version:100
% Created Tuesday March 28, 2006
% This is Copyrighted Material
% For Copying permissions see COPYING.m
% Comments? e-mail sparselab@stanford.edu

az = abs(z);
[sz,iz] = sort(az);
pobs = erfc(sz./sqrt(2));
N = 1:length(z);
pnull =  N' ./length(z);
good = (reverse(pobs) <= (q .* pnull));
if any(good),
    izmax  = iz(1 + length(sz) - max(N(good)));
    thresh = az(izmax);
else
    thresh = max(az) + .01;
end
```

6

Joint Sparsity

In the prior chapters, we studied solving sparse inverse problems of the form

$$y = Ax + n \tag{6.1}$$

The symbols have their usual meaning. This is a single measurement vector (SMV) recovery problem; there is only one measurement "y." In many cases, multiple shots/measurements are acquired.

For example, grayscale image is an SMV problem. But consider the problem of color or hyper-spectral imaging. Here, the same scene is acquired at multiple wavelengths; this turns out to be a multiple measurement vector (MMV) problem.

Another example can be from medical imaging. Traditionally, in magnetic resonance imaging (MRI), a single receiver channel is used. But in modern scanners, multiple channels are being used to accelerate the scan. The former is an SMV problem, while the latter is an MMV problem.

MMV problems can also arise in areas such as dynamic MRI or dynamic computed tomography (CT). Here, the different frames, collected over time, form the MMVs.

In general, MMV inverse problems are expressed as,

$$Y = AX + N, \text{ where } Y = [y_1 | \ldots | y_N] \text{ and } X = [x_1 | \ldots | x_N] \tag{6.2}$$

In this chapter, we are interested in a particular class of MMV problems, where the solutions are row-sparse or joint-sparse. In such cases, the recovery is expressed as,

$$\min_X \|X\|_{2,0} \text{ subject to } \|Y - AX\|_F^2 \leq \varepsilon \tag{6.3}$$

Here, the $l_{2,0}$-norm is a mixed norm; it is defined as the number of l_2-norms of the rows. Let us take a moment to understand this. The "number of rows" acts as an outer l_1-norm; it enforces minimizing the number of selected rows, thus enforcing row sparsity. The inner l_2-norm on the rows promotes density within the row; that is, in the selected row, all the elements will be non-zeroes. In this fashion, the combined $l_{2,0}$-norm enforces row sparsity/joint sparsity.

Just like l_0-minimization, $l_{2,0}$-minimization is non-deterministic polynomial (NP) hard. Hence, one either needs to relax it or solve it heuristically via greedy techniques. In this chapter, we will first study relaxation-based convex optimization methods and then greedy techniques.

Convex Relaxation

The formulation (6.3) is a constrained formulation. This is preferable by the signal processing community. Since in many cases, one can characterize the noise, ε is can be estimated. However, so far, we have learnt to solve only unconstrained problems. For example, the unconstrained relaxation of (6.3) is,

$$\min_X \|Y - AX\|_F^2 + \lambda \|X\|_{2,1} \tag{6.4}$$

The $l_{2,1}$-norm is the convex relaxation of the $l_{2,0}$-norm. It is defined as the sum of the l_2-norm of the rows. Just as the l_1-norm relaxes the "number of coefficients" by its absolute sum, the $l_{2,1}$-norm relaxes the "number of rows" by the sum of their l_2-norms.

The constrained (6.3) and the unconstrained (6.4) formulations are equivalent for proper choice of the Lagrangian λ. Unfortunately, for most practical problems, it is not possible to determine λ explicitly by analytical means. Therefore, instead of "guessing" λ, given the value of ε, we will reach the solution of the constrained problem by iteratively solving a series of unconstrained problems with decreasing values of λ. Such cooling techniques are successful, since the Pareto curve for the said problem is smooth.

The formulations (6.3) and (6.4) are the synthesis prior problems. One can also have analysis prior problems. In general, we will learn to solve the following problems by using the majorization-minimization (MM) approach,

$$\text{Synthesis prior: } \min_X \|X\|_{2,1} \text{ subject to } \|Y - AX\|_F^2 \le \varepsilon \tag{6.5}$$

$$\text{Analysis prior: } \min_X \|HX\|_{2,1} \text{ subject to } \|Y - AX\|_F^2 \le \varepsilon \tag{6.6}$$

Instead of solving the aforesaid constrained problems, we propose solving their unconstrained counterparts,

$$\min_X J_1(X), \text{ where } J_1(X) = \frac{1}{2}\|Y - AX\|_F^2 + \lambda\|X\|_{2,1} \tag{6.7}$$

$$\min_X J_2(X), \text{ where } J_2(X) = \frac{1}{2}\|Y - AX\|_F^2 + \lambda\|HX\|_{2,1} \tag{6.8}$$

Majorization Minimization

We have used this technique several times before; so, we skip the details of the derivation. MM of (6.7) and (6.8) leads to

$$\min_X G_1^{(i)}(X),\ G_1^{(i)}(X) = \frac{1}{2}\|B^{(i)} - X\|_F^2 + \frac{\lambda}{\alpha}\|X\|_{2,1} \tag{6.9}$$

$$\min_X G_2^{(i)}(X),\ G_2^{(i)}(X) = \frac{1}{2}\|B^{(i)} - X\|_F^2 + \frac{\lambda}{\alpha}\|HX\|_{2,1} \tag{6.10}$$

where:

$$B^{(i)} = X^{(i)} + \frac{1}{a}A^T(Y - AX^{(i)}) \tag{6.11}$$

Here, a is greater than the maximum eigenvalue of A^TA.

Solving the Synthesis Prior Problem

For the synthesis prior problem, we need to solve (6.9) at each iteration. Taking the derivative of $G_1^{(i)}(X)$, we get,

$$\frac{dG_1^{(i)}(X)}{dX} = X - B^{(i)} + \frac{\lambda}{a}\Lambda signum(X) \tag{6.12}$$

where $\Lambda = diag(\|X^{(i)j\to}\|_2^{-1})|X^{(i)}|$.

Setting the derivative to zero and re-arranging, we get,

$$B = X + \frac{\lambda}{a}\Lambda signum(X) \tag{6.13}$$

This can be solved by the following soft thresholding,

$$X^{(i+1)} = signum(B^{(i)})\max(0,\left|B^{(i)}\right| - \frac{\lambda}{a}\Lambda) \tag{6.14}$$

Equations (6.11) and (6.14) suggest a compact solution for the unconstrained synthesis prior problem. This is given in the following algorithm.

Initialize: $X^{(0)} = 0$
Repeat until convergence (for iteration i):
Landweber iteration: $B^{(i)} = X^{(i)} + 1/\alpha\, H^T(Y - HX^{(i)})$
Modified soft thresholding: $X^{(i+1)} = \text{signum}(B^{(i)})\max(0,\left|B^{(i)}\right| - \lambda/\alpha\,\Lambda)$

Solving the Analysis Prior Problem

Solving the analysis prior problem requires minimization of (6.10) in each iteration. Taking the derivative of $G_2^{(i)}(X)$, we get,

$$\frac{dG_2^{(i)}(X)}{dX} = X - B^{(i)} + \frac{\lambda}{\alpha}H^T\Omega HX \tag{6.15}$$

where $\Omega = diag(\left\|w^{(i)j\rightarrow}\right\|_2^{-1})$ and $W_{M\times r} = H_{M\times N}X_{N\times r}$.

Setting the gradient to zero, we get,

$$\left(I + \frac{\lambda}{a}H^T\Omega H\right)X = B^{(i)} \tag{6.16}$$

It is not possible to solve (6.16) directly, as the sparsifying transform (H) in most cases is available as a fast operator and not as an explicit matrix. The derivation of the solution to (6.16) is similar to the derivation of analysis prior sparse optimization and analysis prior group-sparse optimization. Thus, we skip the detailed derivations and show the final update equations,

$$Z^{(i+1)} = (\frac{a}{\lambda}\Omega^{-1} + cI)^{-1}(cZ^{(i)} + H(B^{(i)} - H^T Z^{(i)}))$$

$$X^{(i+1)} = B^{(i)} - H^T Z^{(i)} \tag{6.17}$$

where c is greater than the maximum eigenvalue of $A^T A$.

This leads to the following algorithm for solving the analysis prior joint-sparse optimization problem.

Initialize: $X^{(0)} = 0$
Repeat until convergence (in iteration i):
Landweber iteration: $B^{(i)} = X^{(i)} + 1/\alpha\, A^T(Y - AX^{(i)})$
Update auxiliary variable: $Z^{(i+1)} = (\alpha/\lambda\,\Omega^{-1} + cI)^{-1}(cZ^{(i)} + H(B^{(i)} - H^T Z^{(i)}))$
Update variable: $X^{(i+1)} = B^{(i)} - H^T Z^{(i)}$

Solving the Constrained Problem via Cooling

We have derived algorithms to solve the unconstrained problems. As mentioned earlier, the constrained and unconstrained forms are equivalent for proper choice of ε and λ. However, there is no analytical relationship between them in general. The cooling technique solves the constrained problem in two loops. The outer loop decreases the value of λ. The inner loop solves the unconstrained problem for a specific value λ. As λ progressively decreases, the solution of the unconstrained problem reaches the desired solution. Such a cooling technique works because the Pareto curve between the objective function and the constraint is smooth. The cooling algorithms for the synthesis prior and the analysis prior are as follows:

SYNTHESIS PRIOR ALGORITHM

Initialize: $X^{(0)} = 0; \lambda < \max(A^T x)$
Choose a decrease factor (*DecFac*) for cooling λ
Outer Loop: While[1] $\|y - Ax\|_F^2 \geq \varepsilon$
Inner Loop: While[2] $J^{(i)} - J^{(i+1)}/J^{(i)} + J^{(i+1)} \geq Tol$
$J^{(i)} = \|Y - AX^{(i)}\|_F^2 + \lambda\, \|X^{(i)}\|_{2,1}$
Compute: $B^{(i)} = X^{(i)} + 1/a\, A^T(Y - AX^{(i)})$
Compute: $X^{(i+1)} = signum(B^{(i)})\max(0, \|B^{(i)}\| - \lambda/\alpha\, \Lambda)$
$J^{(i+1)} = \|Y - AX^{(i)}\|_F^2 + \lambda\, \|X^{(i)}\|_{2,1}$
End While[2] (inner loop ends)
Cool: $\lambda = \lambda \times DecFac$
End While[1] (outer loop ends)

ANALYSIS PRIOR ALGORITHM

Initialize: $X^{(0)}=0$; $\lambda < \max(A^T x)$
Choose a decrease factor (*DecFac*) for cooling λ
Outer Loop: While[1] $\| y - Ax \|_F^2 \geq \varepsilon$
Inner Loop: While[2] $J^{(i)} - J^{(i+1)} / J^{(i)} + J^{(i+1)} \geq Tol$
$J^{(i)} = \left\| Y - AX^{(i)} \right\|_F^2 + \lambda \left\| HX^{(i)} \right\|_{2,1}$
Compute: $B^{(i)} = X^{(i)} + 1/a\, A^T (Y - AX^{(i)})$
Update: $Z^{(i+1)} = (a/\lambda \Omega^{-1} + cI)^{-1} (cZ^{(i)} + H(B^{(i)} - H^T Z^{(i)}))$
Update: $X^{(i+1)} = B^{(i)} - H^T Z^{(i)}$
$J^{(i+1)} = \left\| Y - AX^{(i)} \right\|_F^2 + \lambda \left\| HX^{(i)} \right\|_{2,1}$
End While[2] (inner loop ends)
Cool: $\lambda = \lambda \times DecFac$
End While[1] (outer loop ends)

Greedy Methods

Just as there are greedy methods to solve the sparse and group-sparse recovery problems, there are techniques to solve the row-sparse recovery problem as well. It is a simple extension of the sparse recovery for SMV. Let us recall the algorithm for orthogonal matching pursuit (OMP), used for solving the SMV problem (6.18).

$$y = Ax \tag{6.18}$$

ORTHOGONAL MATCHING PURSUIT

Initialize: $\Lambda = \{\}$ and $x = 0$.
Until convergence repeat:
Compute: $c = A^T (y - Ax^{(k)})$
Detect support: $l^{(k)} = \arg\max |c_j|$
Append to existing support: $\Lambda = \Lambda \cup l$
Estimate solution: $x^{(k+1)} = \arg\min \left\| y - A_\Lambda x_\Lambda \right\|_2^2$

In the first stem, the correlation between the columns of A and the residual from the previous iteration is computed. The residual is orthogonal to the current solution. In the support detection step, the position of the highest correlate is noted; in the support update stage, the detected support is appended to the existing support. Only those columns of A in the support set are used for solving the coefficients of the solution at the current support. The reasoning behind these steps have been discussed at length in a previous chapter and hence will not be repeated here.

For the given problem, we need to find a solution that is row-sparse. This is done in simultaneous orthogonal matching pursuit (SOMP). The only change in the algorithm will be in the criterion for support detection. Instead of choosing a single coefficient, we have to choose a row. This is easily done by looking at the correlate. For the current problem, it will be a matrix $C = A^T Y$. In OMP, the position of highest absolute correlate was chosen as the support; in this case, we compute the l2-norm of each row of C and choose the row with the highest l2-norm as the support. The rest of the algorithm remains as it is.

Note that for an SMV problem, the same algorithm (computing l2-norm of rows) will boil down to computing the absolute value of correlate. Hence, the SOMP will become the same as OMP.

SIMULTANEOUS ORTHOGONAL MATCHING PURSUIT

Initialize: $\Lambda=\{\}$ and $X = 0$.
Until convergence repeat:
Compute: $C = A^T(Y - AX^{(k)})$
Detect support: $l^{(k)} = \arg\max \|C^{j\rightarrow}\|_2$
Append to existing support: $\Lambda = \Lambda \cup l$
Estimate solution: $X^{(k+1)} = \arg\min \|Y - A_\Lambda X_\Lambda\|_F^2$

Just as there are different greedy approaches for the sparse and group-sparse recovery problems, there can be different variants for the row-sparse recovery problem as well. We are not going to discuss these, but one example is given in MATLAB® codes.

Appendix: MATLAB® Codes

Synthesis MMV Recovery

```
function X = SynthMMV(Y, HforVec, err, normfac)

% Solves problem of the form:
% min ||X||_2,1 s.t. ||Y-HX|| < err

% Inputs
% Y - n x r observation vectors
% H - n x N projection operator
% err - error tolerance
% normfac - maximum eigenvalue of projection operator

% Outputs
% X - N x r input vectors to be recovered

% Converting input matrix to operator
explicitA = ~(ischar(HforVec) || isa(HforVec, 'function_
handle'));
if (explicitA)
    HOp = opMatrix(HforVec); clear HforVec
else
    HOp = HforVec;
end

r = size(Y,2); % number of observations/inputs

H = opMatWrapper(HOp, r); % wrapper for handling matrix
inputs

alpha = 1.05*(normfac^2);

X = H(Y,2); % Initialize X
N = size(X,1); % length of each input vector

maxIter = 100; % Define the maximum number of iterations
tol = 1e-4; % tolerance level
decfac = 0.5; % decrease factor for lambda

lambdaInit = decfac*max(max((abs(X))));
lambda = lambdaInit;
```

```
while lambda > lambdaInit*tol
    iter = 0;
    while iter < maxIter
        iter = iter + 1;
        for i = 1:N
            D(i,:) = (1/norm(X(i,:))).*abs(X(i,:));
        end

        B = X + (1/alpha)*H(Y-H(X,1),2);

        Xvec = SoftTh(B(:),(lambda/alpha).*D(:));
        X = reshape(Xvec,[N r]);

    end
    if norm(Y-H(X,1),'fro') < err
        break;
    end
    lambda = decfac*lambda;
end

    function  z = SoftTh(s,thld)
        z = sign(s).*max(0,abs(s)-thld);
    end
end
```

Analysis MMV Recovery

```
function X = AnaMMV(Y, HforVec, AforVec, err, Hnormfac,
Anormfac)

% Solves problem of the form:
% min ||AX||_2,1 s.t. ||Y-HX|| < err

% Inputs
% Y - n x r observation vectors
% Hforvec - n x N measurement operator
% Aforvec - M X N sparsifyng operator
% err - error tolerance
% Hnormfac - maximum eigenvalue of measurement operator
% Anormfac - maximum eigenvalue of sparsifying transform

% Outputs
% X - N x r input vectors to be recovered
```

```
% Converting input measurement matrix to operator
explicitH = ~(ischar(HforVec) || isa(HforVec,
'function_handle'));
if (explicitH)
    HOp = opMatrix(HforVec); clear HforVec
else
    HOp = HforVec;
end
% Converting input sparsifying matrix to operator
explicitA = ~(ischar(AforVec) || isa(AforVec,
'function_handle'));
if (explicitA)
    AOp = opMatrix(AforVec); clear HforVec
else
    AOp = AforVec;
end

r = size(Y,2); % number of observations/inputs

H = opMatWrapper(HOp, r); % wrapper for handling matrix inputs
A = opMatWrapper(AOp, r); % wrapper for handling matrix inputs

alpha = 1.05*(Hnormfac^2);
c = 1.05*(Anormfac^2);

X = H(Y,2); % Initialize X
Z = A(X,1).*0; % Initialize Z
N = size(X,1); % length of each input vector
M = size(Z,1); % length of transform domain sparse vector

maxIter = 100; % Define the maximum number of iterations
tol = 1e-4; % tolerance level
decfac = 0.5; % decrease factor for lambda

lambdaInit = decfac*max(max((abs(X)))); lambda = lambdaInit;

while lambda > lambdaInit*tol
    iter = 0;
    while iter < maxIter
        iter = iter + 1;

        W = A(X,1);
        for i = 1:M
            D(i,:) = (1/norm(W(i,:)));
        end
```

```
        B = X + (1/alpha)*H(Y-H(X,1),2);
        Z = diag(1./((alpha/lambda)*(1./D) + c))*(c*Z + A(B
        - A(Z,2),1));
        X = B - A(Z,2);

        if norm(Y-H(X,1),'fro') < err
            break;
        end
    end
    lambda = decfac*lambda;
end
end
```

Required Sparco Wrapper

```
function op = opMatWrapper(operator, n)
% OPDCT  One-dimensional discrete cosine transform (DCT)
%
%    OPDCT(N) creates a one-dimensional discrete cosine
transform
%    operator for vectors of length N.

%   Copyright 2008, Ewout van den Berg
and Michael P. Friedlander
%   http://www.cs.ubc.ca/labs/scl/sparco
%   $Id: opDCT.m 1040 2008-06-26 20:29:02Z ewout78 $

op = @(x,mode) opMatWrapper_intrnl(operator,n,x,mode);


function y = opMatWrapper_intrnl(operator,n,x,mode)

if mode == 0
   t = operator([],0);
   y = {t{1},t{2},t{3},{'Matrix Input Matrix Output'}};
elseif mode == 1
    for i = 1:n
        y(:,i) = operator(x(:,i),1);
    end
else
   for i = 1:n
        y(:,i) = operator(x(:,i),2);
    end
end
```

MMV SwOMP

```
function X = MMVSWOMP(Y, A, k)

n = size(A,2); %Get second dimension of A
m = size(Y,2); %Get second dimension of Y

x = zeros(n,m); %This is the dimension of X.
omega = [];
Aomega = A;
Xomega = x;

lambda = 0.7;

for i = 1:5

    r = Y - Aomega*Xomega; %Calculate residuals with the A
and calculated X.

    if ( r < 0.2)   %Check if residual is around the
tolerance value, break.
        break;
    end

    c = abs(A'*r);      %Get correlation of A with r.
    N = sqrt(sum(abs(c).^2,2)); %Calculate row-
wise l2 norm

    ndx = find(N >= lambda*max(N)); %Find indices having
more than or equal to the max row norm(*lambda)

    omega = [omega; ndx]; %Append indices found
    Aomega = A(:,omega);
    Xomega = Aomega\Y; %Calculate Xomega with these indices

    if (size(omega) >= k)   %If k rows have been recovered,
exit.
        break;
    end

end

X = zeros(n,m);
X(omega,:) = Xomega; %Generate the entire matrix X and return.

end
```

7

Low-Rank Matrix Recovery

There is a matrix $X_{n \times n}$ (it can be rectangular as well), but all the entries $(x_{i,j}, (i,j) = 1...n)$ are not available. Only a subset (Ω) of entries is observed. Now, the question is, is it possible to estimate all the entries of the matrix, given the set of partially observed samples? In general, the answer is NO. However, in a special situation, when the matrix is of low rank, it is possible to estimate the entire matrix, provided "enough" samples are available.

The intuitive reason behind this argument is that if the rank of the matrix $r < n$ is low, then the number of degrees of freedom is only $r(2n - r)$. When the rank (r) is small, the number of degrees of freedom is considerably less than the total number of entries ($n \times n$) in the matrix. Therefore, we can hope that there might be a possibility that all entries of the matrix are recoverable from a subset of it.

The above discussion can be framed in the following optimization problem,

$$\begin{aligned} &\min \, rank(X) \\ &\text{subject to } Y = M_\Omega(X) \end{aligned} \tag{7.1}$$

where, M_Ω is the masking operator that selects the entries in X falling in the set Ω, and Y are the available samples.

Unfortunately, (7.1) is an non-deterministic polynomial (NP) hard problem; the complexity of solving this problem is doubly exponential.

Researchers in machine learning and signal processing have been interested in this problem in the past few years for a variety of problems: collaborative filtering, system identification, global positioning, direction of arrival estimation, etc. What they did was to solve the tightest convex surrogate of rank minimization, which is the following,

$$\begin{aligned} &\min \, \|X\|_* \\ &\text{subject to } Y = M_\Omega(X) \end{aligned} \tag{7.2}$$

where $\|X\|_*$ is the nuclear norm (also called Trace norm or Ky-Fan norm) of the matrix and is defined as the sum of absolute singular values of the matrix.

The nuclear norm is the closest (tightest) convex relaxation to the NP hard rank minimization problem. Therefore, solving (7.2) and expecting a result similar to (7.1) were in some way expected.

In recent times, mathematicians are taking a closer look at the matrix completion problem and its relation with nuclear norm minimization. They are mainly interested in finding the bounds on the number of entries required to estimate the complete matrix and the conditions necessary to be imposed on the nature of the matrix and the mask (sampling strategy). We will briefly discuss these theoretical results.

The fundamental theorem behind the matrix completion problem states the following: Suppose we observe m entries of X, with locations sampled uniformly at random. Then, there is a positive numerical constant C, such that if

$$m \geq C\mu^a nr \log^b n, \tag{7.3}$$

the nuclear norm minimization problem (7.2) will find the unique solution with probability $1-n^{-d}$.

Here, μ depends on the nature of X (will be discussed shortly).

The integer constants a, b, and d vary depending on the approach of the proof. It is important to note that the theorem proves that an NP hard problem (7.1) can be solved by a convex relaxation (7.2), with somewhat larger number of samples with a very high probability.

Now, consider a pathological low-rank matrix that has all entries as zeros except for a single non-zero value. Even though the matrix is of rank 1, it is impossible to recover the matrix without sampling all the entries. The same is true for a matrix that has only one row/column of non-zero entries, while the rest are zeroes. The behavior of such matrices can be explained by understanding the factor μ.

Let $X = U\Sigma V^T$ be the singular value decomposition (SVD) of the matrix under consideration, where $U = [u_1,...,u_r]$ and $V = [v_1,...,v_r]$ be the right and left singular vectors, respectively, and $\Sigma = diag(\sigma_1,...,\sigma_r)$ be the singular values. Now, μ is defined as,

$$\max(u_k) \leq \sqrt{\mu / n} \text{ and } \max(v_k) \leq \sqrt{\mu / n} \tag{7.4}$$

where max(w) is the maximum absolute value in the vector w.

Expressed in words, the singular vectors should not be too spiky. If they are spiky, then the μ is high, and therefore, the number of samples needed for perfect recovery (7.3) is also high. This is called the "incoherence property." For the pathological matrices discussed earlier, the singular values are spiky and therefore would require lot of samples to estimate the matrix. This explains the pathological matrices mentioned earlier. In those cases, the singular values would be spiky and the matrix will have non-zero value at one position and zeroes everywhere else. This would make the value of μ to be high, which in turn would indicate that a very large number of samples needs to be collected in that case.

Lastly, we will discuss why the theorem has been proved for samples "collected uniformly at random". If there are no entries selected from a particular row or column, then it is impossible to reconstruct that row or column for even a matrix of rank unity. When the samples are collected uniformly at random, the number of samples required to ensure that every row and column is sampled at least once is O(nlogn); this is as good an estimate one can get for a problem of size n^2.

Here, we discussed about the problem of matrix completion of a low-rank matrix; this has been generalized. Instead of using a masking operator, one can use any linear measurement operator. In cases where the number of measurements is less than the size of the matrix, the low-rank matrix can be recovered by nuclear norm minimization.

Connections with Compressed Sensing

Compressed sensing studies the problem of solving a system of under-determined linear equations when the solution is known to be sparse. Consider the following system of equations,

$$y_{m\times 1} = M_{m\times n}x_{n\times 1},\ m < n \tag{7.5}$$

In general, (7.5) has infinitely many solutions. But if the solution is known to be sparse, it has been proved that the solution is necessarily unique. Assume that the vector x is k-sparse; that is, it has k non-zero entries, while the rest are all zeroes. In such a scenario, there are only $2k$ (k-positions and k-values) unknowns. Now, as long as the number of equations $m \geq 2k$, it is possible to solve (7.5). Mathematically, the problem can be stated as,

$$\begin{gathered}\min \|x\|_0 \\ \text{subject to } y = Mx\end{gathered} \tag{7.6}$$

$\|x\|_0$ is not a norm in the strictest sense; it only counts the number of non-zero entries in the vector.

In words, (7.6) means that, of all the possible solutions (7.5), chose the sparsest one.

Unfortunately, solving (7.6) is known to be an NP hard problem [11]. There is no known algorithm that has shown any significant improvement compared with brute force solution of (7.5). In machine learning and signal processing, instead of solving the NP hard l_0-norm minimization problem, its closest convex surrogate (l_1-norm minimization problem) is generally solved,

$$\min \|x\|_1 \quad \text{subject to } y = Mx \tag{7.7}$$

It is known that for several types of matrices (M), solving (7.6) and (7.7) is equivalent, that is, both of them yielded the sparsest solution.

Solving (7.7) is easy, since it is a convex problem and can be solved by linear programming. But, the number of equations required to solve (7.5) via convex optimization (7.7) is significantly more than that required to solve the NP hard problem (7.6). The number of equations required also depends on the type of matrix (M); for some common matrices, the number of equations required is,

$$m \geq Ck\log(n) \tag{7.8}$$

The trade-off is expected; the ease of solving the inverse problem comes at the cost of larger number of equations!

Both compressed sensing (CS) and matrix completion (MC) study problems where the number of unknowns is apparently larger than the number of equations. In CS, the length of the vector to be estimated is larger than the number of equations; in MC, only a subset of the elements of the matrix is known. In general, none of the problems has a unique solution. However, when the degrees of freedom in the solution are less than the number of equations, both problems have a unique solution. For CS, the degree of freedom is $2k$, where k is the number of non-zero elements in the vector; for MC, the degree of freedom is $r(2n-r)$, where r is the rank of the matrix and n^2 is the number of elements in the matrix.

FOCUSS Algorithm

The problem we intend to solve is of the form

$$\min \|X\|_* \text{ subject to } y = Ax, x = vec(X) \tag{7.9}$$

For employing the FOCally Underdetermined System Solver (FOCUSS) approach, we will define the nuclear norm of the matrix X in terms of its Ky-Fan norm, that is, $\|X\|_* = Tr(X^T X)^{1/2}$; therefore, the problem to be solved is,

$$\min Tr(X^T X)^{1/2} \text{ subject to } y = Ax,\ x = vec(X) \tag{7.10}$$

The unconstrained Lagrangian form for (7.10) is,

$$L(x,\lambda) = Tr(X^T X)^{1/2} + \lambda^T (y - Ax) \tag{7.11}$$

where λ is the vector of Lagrangian multipliers.

The conditions for stationary points of (7.11) are,

$$\nabla_x L(X,\lambda) = (XX^T)^{-\frac{1}{2}} X + A^T \lambda = 0 \tag{7.12}$$

$$\nabla_\lambda L(X,\lambda) = Ax - y = 0 \tag{7.13}$$

Now, (7.9) can be expressed as,

$$pWx + A^T \lambda = 0,\ W = I \otimes (XX^T)^{-\frac{1}{2}} \tag{7.14}$$

where $\otimes$ denotes the Kronecker product.

Solving x from (7.14),

$$x = -W^{-1} A^T \lambda \tag{7.15}$$

W is a block diagonal matrix with positive semi-definite blocks along the diagonal. The problem is that, since W is positive semi-definite, the solution is not numerically stable. Such a problem was encountered while using FOCUSS for sparse signal recovery in compressed sensing; in order to get a stable solution, W must be positive definite and hence must be regularized.

$$W_k = I \otimes (X_{k-1} X_{k-1}^T + \varepsilon I)^{-\frac{1}{2}} \tag{7.16}$$

Here, ε is a small constant that regularizes the solution. This regularization also guarantees that W (and hence W^{-1}) is positive definite. As $\varepsilon \to 0$, one arrives at the desired solution.

Substituting the value of x from (7.15) into (7.13) and solving for λ, we get,

$$\lambda = -(AW^{-1}A^T)^{-1}y \tag{7.17}$$

Substituting the value of λ back in (7.15), we get,

$$x = W^{-1}A^T(AW^{-1}A^T)^{-1}y \tag{7.18}$$

In order to efficiently compute x in each iteration, we re-write (7.18) as,

$$x = R\tilde{x}, \text{ where } \tilde{x} = (AR)^T((AR)(AR)^T)^{-1}y \tag{7.19}$$

Here, R is the Cholesky decomposition of W^{-1}. The decomposition exists since W^{-1} is a positive definite matrix. The reason we expressed (7.18) in the current form (7.19) is because $\tilde{x}$ can be solved very efficiently by using the LSQR algorithm. Based on this modification, we propose the following efficient algorithm to solve the nuclear norm minimization problem.

> Intitialize: $x_0 = A^T(AA^T)^{-1}y$, which is a least squares solution; define ε
> Repeat until stopping criterion is met:
> Compute: $W_k = I \otimes (X_{k-1}X_{k-1}^T + \varepsilon I)^{-\frac{1}{2}}$ and $R_k R_k^T = W_k^{-1}$.
> Update: $\tilde{x}_k = (AR_k)^T((AR_k)(AR_k)^T)^{-1}y$ and $x_k = R\tilde{x}_k$. Reshape x_k to matrix form X_k.
> Decrease: $\varepsilon = \varepsilon / 10$ iff $\|x_k - x_{k-1}\|_2 \le tol$

We use two stopping criteria. The first one limits the maximum number of iterations to a certain value. The second stopping criterion is based on the value of change in the objective function in two consecutive iterations; therefore, if the change is nominal, then the iterations stop assuming that the solution has reached a local minimum. The update step of the algorithm is solved by LSQR, which runs for a maximum of 20 iterations. The value of ε is initialized at 1. The tolerance level for deciding the decrease of ε is fixed at 10^{-3}.

Singular Value Shrinkage

The FOCUSS method is efficient and gives good results but does not directly account for the fact that the retrieved matrix is low-rank. Also, it is applicable only for the noise-free scenario. In realistic cases, where there is noise, we would better be solving

$$\min \|y - Ax\|_2^2 + \lambda \|X\|_*, x = vec(X) \tag{7.20}$$

We use the majorization-minimization (MM) technique to solve it. Since we have used it many times in the past, we will not repeat the intermediate steps here. We will directly go to the steps that we need to solve in every iteration (*k*).

$$\min \|b - x\|_2^2 + \lambda \|X\|_* \tag{7.21}$$

where $b = x_{k-1} + 1/\alpha\, A^T(y - Ax_{k-1})$; α is the maximum eigenvalue of A^TA.

Now, x and b are vectorized forms of matrices. The following property of singular value decomposition holds in general,

$$\|A_1 - A_2\|_F^2 \geq \|\Sigma_1 - \Sigma_2\|_F^2 \tag{7.22}$$

where A_1 and A_2 denote two matrices and Σ_1 and Σ_2 are their singular value matrices, respectively.

Using this property, minimizing (7.21) is the same as minimizing the following,

$$\min_s \|s - s_{k+1}\|_2^2 + \frac{\lambda}{\alpha}\|s\|_1 \tag{7.23}$$

where s and s_{k+1} are the singular values of matrices corresponding to x and b, respectively.

It is possible to write (7.23) in a decoupled fashion,

$$\min_s \|s - s_{k+1}\|_2^2 + \frac{\lambda}{\alpha}\|s\|_1 = (s^{(1)} - s_{k+1}^{(1)})^2 + \frac{\lambda}{\alpha}|s^{(1)}| + \ldots + (s^{(r)} - s_{k+1}^{(r)})^2 + \frac{\lambda}{\alpha}|s^{(r)}| \tag{7.24}$$

Therefore, (7.24) can be minimized by minimizing each of the terms,

$$f(s^{(i)}) = (s^{(i)} - s_{k+1}^{(i)})^2 + \frac{\lambda}{\alpha}|s^{(i)}| \tag{7.25}$$

$$\frac{\partial f(s^{(i)})}{\partial s^{(i)}} = 2(s^{(i)} - s_{k+1}^{i}) + \frac{\lambda}{\alpha} signum(s^{(i)})$$

Setting the derivative to zero and rearranging, we get

$$s_{k+1}^{(i)} = s^{(i)} + \frac{\lambda}{2\alpha} signum(s^{(i)}) \tag{7.26}$$

We already know that the function that minimizes (7.26) is the following,

$$s^{(i)} = signum(s_{k+1}^{i})\max\left(0, \left|s_{k+1}^{i}\right| - \frac{\lambda}{2\alpha}\right)$$

Written for the entire vector of singular values,

$$s = soft\left(s_{k+1}, \frac{\lambda}{2\alpha}\right) = signum(s_{k+1})\max\left(0, \left|s_{k+1}\right| - \frac{\lambda}{2\alpha}\right) \tag{7.27}$$

This is the famous soft thresholding function used profusely in sparse.

Equations (7.18) and (7.27) suggest the shrinkage algorithm:

SHRINKAGE ALGORITHM

1. $b = x_{k-1} + 1/\alpha A^T(y - Ax_{k-1})$
2. Form the matrix B by reshaping b.
3. SVD: $B = U\Sigma V^T$.
4. Soft threshold the singular values: $\hat{\Sigma} = soft(diag(\Sigma), \lambda/2\alpha)$
5. Update variable: $X_{k+1} = U\hat{\Sigma}V^T$. Form x_{k+1} by vectorizing X_{k+1}.
6. Update: $k = k + 1$ and return to step 1.

Singular Value Hard Thresholding

The rank minimization problem can be expressed as,

$$\min \left\|y - Ax\right\|_2^2 + \lambda \text{rank}(X) \tag{7.28}$$

This is an NP hard problem; so, we cannot solve it exactly. However, we can solve it approximately. Using the same MM technique, we arrive at the following problem in every iteration,

$$\min_s \left\|s - s_{k+1}\right\|_2^2 + \frac{\lambda}{\alpha}\left\|s\right\|_0 \tag{7.29}$$

Note that, since it is a "rank minimization" problem, we have the l_0-norm on the singular values instead of the l_1-norm (as used in nuclear norm).

Equation (7.29) decouples the vectors element-wise, and therefore, we have,

$$\|s - s_{k+1}\|_2^2 + \frac{\lambda}{\alpha}\|s\|_0 = (s^{(1)} - s_{k+1}^{(1)})^2 + \frac{\lambda}{\alpha}\left|s^{(1)}\right|^0 + \ldots + (s^{(r)} - s_{k+1}^{(r)})^2 + \frac{\lambda}{\alpha}\left|s^{(r)}\right|^0 \quad (7.30)$$

Since the optimization problem is decoupled, one can proceed to minimize (7.30) element-wise. We now follow an analysis similar to iterative hard thresholding in sparse recovery. To derive the minimum, two cases need to be considered: case 1: $s^{(i)} = 0$ and case 2: $s^{(i)} \neq 0$. The element-wise cost is 0 in the first case. For the second case, the cost is $(s^{(i)})^2 - 2s^{(i)}s_{k+1}^{(i)} + \lambda/\alpha$, the minimum of which is reached when $s^{(i)} = s_{k+1}^{(i)}$.

Comparing the cost in both cases, that is,

$$0 \text{ if } s^{(i)} = 0$$

$$-(s_{k+1}^{(i)})^2 + \frac{\lambda}{\alpha} \text{ if } s^{(i)} = s_{k+1}^{(i)}$$

We see that the minimum of each decoupled term in (7.30) is achieved when,

$$s^{(i)} = \begin{cases} s_{k+1}^{(i)} \text{ when } s_{k+1}^{(i)} > \lambda / 2\alpha \\ 0 \quad \text{ when } s_{k+1}^{(i)} \leq 0 \end{cases} \quad (7.31)$$

When (7.31) is applied on the whole vector element-wise, it is popularly called hard thresholding, and we can write,

$$s = hard\left(s_{k+1}, \frac{\lambda}{2\alpha}\right) \quad (7.32)$$

Based on these equations, we can have a hard-thresholding algorithm similar to the shrinkage algorithm of the previous subsection,

HARD-THRESHOLDING ALGORITHM

1. $b = x_{k-1} + 1/\alpha\, A^T(y - Ax_{k-1})$
2. Form the matrix B by reshaping x_k.
3. SVD: $B = U\Sigma V^T$.
4. Soft threshold the singular values: $\hat{\Sigma} = hard\,(diag(\Sigma), \lambda/2\alpha)$
5. $X_{k+1} = U\hat{\Sigma}V^T$. Form x_{k+1} by vectorizing X_{k+1}.
6. Update: $k = k + 1$ and return to step 1.

Split Bregman Technique

The objective remains the same as before, that is, we want to solve

$$\min \frac{1}{2}\|y - Ax\|_F^2 + \lambda\|X\|_* \tag{7.33}$$

The Split Bregman approach reformulates (7.33) as follows,

$$\min_{X,W} \frac{1}{2}\|y - Az\|_2^2 + \lambda_1\|W\|_* \; Subject\, to\, X = W \tag{7.34}$$

The above formulation can be converted into unconstrained convex optimization problem (7.21) by the use of augmented Lagrangian by introducing a Bregman relaxation variable.

$$\min_{X,W} \left(\frac{1}{2}\right)\|y - Az\|_2^2 + \lambda_1\|W\|_* + \left(\frac{\eta}{2}\right)\|W - X - B\|_2^2 \tag{7.35}$$

where, B is the Bregman (relaxation) variable.

The use of second L2 term (augmented Lagrangian) improves the robustness of the algorithm, as it eliminates the need for strictly reinforcing the equality constraint, while simultaneously penalizing for any deviation. Use of Bregman variable makes sure that the values of λ_1 and η can be chosen to optimize the convergence speed. Hence, the speed of algorithm is dependent on how fast we can optimize each of the subproblems. Equation (7.35) can be split into two simpler subproblems, which can be solved by alternatively fixing one variable and minimizing over the other (via alternating direction method of multipliers (ADMM)).

Subproblem 1:

$$\min_{X} \left(\frac{1}{2}\right)\|y - Ax\|_2^2 + \left(\frac{\eta}{2}\right)\|W - X - B\|_2^2 \tag{7.36}$$

Subproblem 2:

$$\min_{W} \lambda_1\|W\|_* + \left(\frac{\eta}{2}\right)\|W - X - B\|_2^2 \tag{7.37}$$

The first subproblem (7.36) is a simple least squares problem (Tikhonov regularization), which can be solved easily by using any gradient descent algorithm such as conjugate gradient. Subproblem 2 (7.37) can be solved iteratively by soft thresholding operation, as shown in Chapter 3.

$$Soft(t,u) = sign(t)\ \max(0,|t|-u) \tag{7.38}$$

$$W \leftarrow Soft\ (X+B, 4\lambda_1/\eta) \tag{7.39}$$

Between every consecutive iteration of the two subproblems, Bregman variable is updated as follows,

$$B^{k+1} \leftarrow B^k - (W^{k+1} - X^{k+1}) \tag{7.40}$$

Initialization : $B^0 = 1;\ W^0 = 0;\ \eta = 0.001;\ \lambda_1 = 0.001$
while $k < 500$ & $abs(Obj\_func\ (k) - Obj\_func\ (k-1)) < e^{-7}$
$Z^{k+1} = [A^T A + \eta I]^{-1}[\eta(W^k - B^k) + A^T y]$
$W^{k+1} \leftarrow \left(Z^{k+1} + B^k, \frac{4\lambda_1}{\eta}\right)$
$B^{k+1} \leftarrow B^k - \left(W^{k+1} - Z^{k+1}\right)$
end while

There are two stopping criteria. Iterations continue till the objective function converges; by convergence, we mean that the difference between the objective functions between two successive iterations is very small (1E-7). The other stopping criterion is a limit on the maximum number of iterations. We have kept it to be 500.

Matrix Factorization

Nuclear norm minimization is a recent technique, but prior to that, people employed the simpler matrix factorization-based method to recover low-rank matrices. It is not as elegant as nuclear norm minimization but is computationally more efficient.

We are interested in recovering a low-rank matrix X of size m by n. Assuming that we know the rank (r), we can write,

$$X_{m \times n} = U_{m \times r} V_{r \times n} \tag{7.41}$$

This allows us to express the matrix recovery problem as follows,

$$y = A\ vec(X) = A\ vec(UV) \tag{7.42}$$

Since the low-rank matrix is factored into two full-rank matrices, this is called a matrix factorization problem. The recovery is posed as,

$$\min_{U,V} \left\| y - A\, vec(UV) \right\|_F^2 \tag{7.43}$$

This is a bi-linear, and hence non-convex, problem. Thus, there is no guarantee that it will converge to the global minimum. This is the reason that we said the nuclear norm to be a more elegant (convex) approach.

Usually, the matrix factorization problem is regularized.

$$\min_{U,V} \left\| y - A\, vec(UV) \right\|_F^2 + \lambda \left(\left\| U \right\|_F^2 + \left\| V \right\|_F^2 \right) \tag{7.44}$$

The algorithm for solving this is simple. First, we will employ majorization minimization on the cost function. This leads to,

$$\min_{U,V} \left\| b - vec(UV) \right\|_F^2 + \lambda \left(\left\| U \right\|_F^2 + \left\| V \right\|_F^2 \right) \tag{7.45}$$

where $b = vec(UV)_{k-1} + 1/\alpha A^T (y - Avec(UV)_{k-1})$.

This (7.45) can be expressed after reshaping b into a matrix B, having the same dimensions as X.

$$\min_{U,V} \left\| B - UV \right\|_F^2 + \lambda \left(\left\| U \right\|_F^2 + \left\| V \right\|_F^2 \right)$$

This can be solved easily by using alternating minimization. We start with an initialized U, and in every iteration, we update V and U. The initialization of U can be random, but we figured out that a better option is to initialize it deterministically. One can compute the singular value decomposition of X and deterministically initialize U by selecting the top r singular vectors from the left singular vectors. One can also scale these singular vectors by the corresponding singular values.

Initialize: U_0
Iterate till convergence
Update: $V_k \leftarrow \min_V \left\| B - U_{k-1} V \right\|_F^2 + \lambda \left\| V \right\|_F^2$
Update: $U_k \leftarrow \min_U \left\| B - U V_k \right\|_F^2 + \lambda \left\| U \right\|_F^2$

One can run these iterations either for a maximum number of times or till the objective function converges to a local minimum.

Appendix: MATLAB® Codes

FOCUSS Algorithm

```
function [X  bi bu]=
trace_form_nobreg(train,gm,Aop,s_x,lambda_n,iter,lambda_b)
% function [x_out  bi bu  er]= trace_form_nobreg(train,Aop,
s_x,lambda_n,iter)

% In this formulation, nuclear norm is replaced by
equivalent Ky-Fan norm
% This eliminates need for complex singular value
decomposition at every
% iteration and just requires simple least squares
at every step

% We solve
% min_X ||y-A(x+ bu+bi)||_2 + lambda_n||X||_ky-fan +
lambda_b (||bu||_2 +||bi||_2)
% equivalent to
% min_X ||y-A(x+ bu+bi)||_2 + lambda_n[trace{(X'*X)_0.5}]+
lambda_b (||bu||_2 +||bi||_2)

% INPUTS
%train : training set
%Aop : sub-sampling operator
%s_x : Size of data matrix
%lambda_n : regularziation paranmter
%iter : maximum number of iterations

% OUTPUTS
% X : Recovered Matrix

%Initialize variables
X=randn(s_x);
bi=randn(1,s_x(2));
bu=randn(s_x(1),1);
[i j v]=find(train);
train_data=[i j v];
b = Aop(train(:),1);

 for iteration=1:iter

     Bu=repmat(bu,1,s_x(2));
```

```
    Bi=repmat(bi,s_x(1),1);
    bbcs=Bu+Bi+gm;

    % Compute interaction portion
    yvec=b-Aop(bbcs(:),1);
    ynew=X(:)+Aop((yvec-Aop(X(:),1)),2);
    Ymat=reshape(ynew,s_x);

    Xoper=eye(s_x(1)) + lambda_n*((X*X')^(-1/2));
    X=Xoper\Ymat;

    % Compute biases
    [bi, bu]=bias_sgd(train_data,X,bu,bi,gm,lambda_b);

 end

end
```

Singular Value Shrinkage

```
function X = IST_MC(y,M,sizeX,err,x_initial,
normfac,insweep,tol,decfac)

% Matrix Completion via Iterated Soft Thresholding
% min nuclear-norm(X) subject to ||y - M(X)||_2<err

% Inputs
% X - matrix to be estimated
% M - masking operator, applied to vectorized form of X
% y - sampled entries
% err - norm of the mismatch (default 0)
% x_initial - intial estimate of the vectorized
form of X (defalut 0)
% normfac - eigenvalue of (M'M) (default should be 1
for masking operator)
% insweep - maximum number of internal sweeps for
solving ||y - M(X)||_2 + lambda nuclear-norm(X) (default 200)
% tol - tolerance (default 1e-4)
% decfac - decrease factor for cooling lambda

% Copyright (c) Angshul Majumdar 2010

if nargin < 4
    err = 1e-6;
end
```

```
if nargin < 5
    x_initial = zeros(prod(sizeX),1);
end
if nargin < 6
    normfac = 1;
end
if nargin < 7
    insweep = 200;
end
if nargin < 8
    tol = 1e-4;
end
if nargin < 9
    decfac = 0.9;
end

alpha = 1.1*normfac;
x = x_initial;
lambdaInit = decfac*max(abs(M(y,2))); lambda = lambdaInit;
f_current = norm(y-M(x,1)) + lambda*norm(x,1);

while lambda > lambdaInit*tol
    for ins = 1:insweep
        f_previous = f_current;
        x = x + (1/alpha)*M(y - M(x,1),2);
        [U,S,V] = svd(reshape(x,sizeX),'econ');
        s = SoftTh(diag(S),lambda/(2*alpha));
        S = diag(s);
        X = U*S*V';
        x = X(:);
        f_current = norm(y-M(x,1)) + lambda*norm(x,1);
        if norm(f_current-f_previous)/
norm(f_current + f_previous)<tol
            break;
        end
    end
    if norm(y-M(x,1))<err
        break;
    end
    lambda = decfac*lambda;
end

    function  z = SoftTh(s,thld)
        z = sign(s).*max(0,abs(s)-thld);
    end
end
```

Singular Value Hard Thresholding

```
function X = IHT_MC(y,M,sizeX,err,x_initial,normfac,insweep,
tol,decfac)

% Matrix Completion via Iterated Hard Thresholding
% min rank(X) subject to ||y - M(X)||_2<err

% Inputs
% X - matrix to be estimated
% M - masking operator, applied to vectorized form of X
% y - sampled entries
% err - norm of the mismatch (default 0)
% x_initial - intial estimate of the vectorized
form of X (defalut 0)
% normfac - eigenvalue of (M'M) (default should be 1
for masking operator)
% insweep - maximum number of internal sweeps
for solving ||y - M(X)||_2 + lambda rank(X) (default 200)
% tol - tolerance (default 1e-4)
% decfac - decrease factor for cooling lambda

% Copyright (c) Angshul Majumdar 2010

if nargin < 4
    err = 1e-6;
end
if nargin < 5
    x_initial = zeros(prod(sizeX),1);
end
if nargin < 6
    normfac = 1;
end
if nargin < 7
    insweep = 200;
end
if nargin < 8
    tol = 1e-4;
end
if nargin < 9
    decfac = 0.9;
end

alpha = 1.1*normfac;
x = x_initial;
```

```
lambdaInit = decfac*max(abs(M(y,2))); lambda = lambdaInit;
f_current = norm(y-M(x,1)) + lambda*norm(x,1);

while lambda > lambdaInit*tol
    for ins = 1:insweep
        f_previous = f_current;
        x = x + (1/alpha)*M(y - M(x,1),2);
        [U,S,V] = svd(reshape(x,sizeX),'econ');
        s = HardTh(diag(S),lambda/(2*alpha));
        S = diag(s);
        X = U*S*V';
        x = X(:);
        f_current = norm(y-M(x,1)) + lambda*norm(x,1);
        if norm(f_current-f_previous)/
        norm(f_current + f_previous)<tol
            break;
        end
    end
    if norm(y-M(x,1))<err
        break;
    end
    lambda = decfac*lambda;
end

    function  z = HardTh(s,thld)
        z = (abs(s) >= thld) .* s;
    end
end
```

Matrix Factorization

```
function [U, V]=MF(trainset, lambda_u, lambda_v, rnk, max_
iter)

%function [U, V]=MF(xtrain, lambda_u, lambda_v,
rnk, max_iter)

%FUNCTION FOR SOLVING BCS FORMULATION FOR RECOMMENDER SYSTEMS
%minimize_(U,V) ||Y-A(UV)||_2 + lambda_u||U||_F + lambda_
v||V||_F

%EXTRA REQUIREMENT
%Sparco toolbox http://www.cs.ubc.ca/labs/scl/sparco/
```

```
%INPUTS
%Y is the observed rating matrix (training set)
%lambda_u, lambda_v are the regularization parameters
%rnk is the number of latent factors or rank of
rating matrix
%max_iter is the maximum number of iterations

%OUTPUTS
%U is the user latent factor matrix
%V is the item latent factor matrix


%%%%%%%%%%%%%%%%%%%%%%%%%%%%%%%%%%%%%%%%%%%%%%%%%%%%%%%%%%%%%
%Compute size of Rating Matrix
sizeX=size(trainset);

%Find sampled (available rating) points
sampled_pts=find(trainset~=0);
A = opRestriction(prod(sizeX),sampled_pts);

%Compute vector of available ratings
y = A(trainset(:),1);

%Set initial values of output matrices
U = randn(sizeX(1),rnk);
V = rand(rnk,sizeX(2));
x_guess=U*V;

obj_func=[];
%%%%%%%%%%%%%%%%%%%%%%%%%%%%%%%%%%%%%%%%%%%%%%%%%%%%%%%%%%%%%

%%%%%%%%%%%%%%%%%%%%%%%%%%%%%%%%%%%%%%%%%%%%%%%%%%%%%%%%%%%%%
%Begin Iterations

for iter = 1:max_iter

    %%%%%%%%%%%%%%%%%%%%%%%%%%%%%%%%%%%%%%%%%%%%%%%%%%%%%%%%%%%%%
     %Solving subproblem 2
     %minimize_(U) ||Y-A(UV)||_2 + lambda_u||U||_F
```

```
     x_guess=U*V;
    ynew=x_guess(:)+A((y-A(x_guess(:),1)),2);
    Ymat=reshape(ynew,sizeX);

    aa=V*V'+lambda_u*eye(rnk);
    yy=Ymat*V';
    U=yy/aa;
   %%%%%%%%%%%%%%%%%%%%%%%%%%%%%%%%%%%%%%%%%%%%%%%%%%%%%


   %%%%%%%%%%%%%%%%%%%%%%%%%%%%%%%%%%%%%%%%%%%%%%%%%%%%%
    %Solving subproblem 2
    %minimize_(V) ||Y-A(UV)||_2 + lambda_v||V||_1
    x_guess=U*V;
    ynew=x_guess(:)+A((y-A(x_guess(:),1)),2);
    Ymat=reshape(ynew,sizeX);

    ab=U'*U+lambda_v*eye(rnk);
    yb=U'*Ymat;
    V=mldivide(ab,yb);
  %%%%%%%%%%%%%%%%%%%%%%%%%%%%%%%%%%%%%%%%%%%%%%%%%%%%%

  %%%%%%%%%%%%%%%%%%%%%%%%%%%%%%%%%%%%%%%%%%%%%%%%%%%%%
   % Calculate Objective function

   x_guess=U*V;
   data_c = norm((A(x_guess(:),1)-y),2);
   item_m = norm(V(:),2);
   user_m = norm(U(:),2);
   cost = data_c +lambda_u*user_m+lambda_v*item_m;
   obj_func = [obj_func,cost];

   if iter>1
   if  abs((obj_func(iter)-obj_func(iter-1)))<1e-7
       break
   end
   end

end
end
```

Split Bregman

```
function x_out = nuclear_L1_minimize(A,D,y_vec,s_x,lambda1,
lambda2,alpha,c,maxit)
% function x_out = nuclear_L1_minimize(A,D,y_vec,s_x,lambda
1,lambda2,alpha,c,maxit)

% This code solves the problem of recovering a low rank
and sparse(in transform domain)matrix from its
% lower dimensional projections

%Formulated as an unconstarined nuclear norm
and L1 minmization problem
%using Split bregman algorithm, formulation
for the problem is as follows

% Miniimize (lambda1)||W||* + (lambda2)||Dz||_1 + 1/2 |
| A(X) - y ||_2^2 + eta1/2 |
| W-X-B1 ||_2^2 +eta2/2 || W-Z-B2 ||_2^2

%W, Z are auxillary variable
and B1, B2 are the bregman variable

%INPUTS
%A : Linear (sampling) Matrix
%D : Sparifying Transform
%y_vec : Vector of observed values
%s_x : size of the data matrix to
be recovered (in form of [m n])
%lambda1, lambda2 :regularization parameter
%alpha, c : parameters for analysis prior(chambolle)
algorithm
%maxit: Maximum number of iterations

%OUTPUTS
%x_out : recovered low rank and sparse(in transform domain)
matrix

%Algorithm's Internal parameters
eta1=.0001;
eta2=.0001;
gamma=2*alpha/lambda2;

%Initialize variables
s_new=s_x(1)*s_x(2);
W=zeros(s_new,1);
```

```
Z=zeros(s_new,1);
B1=ones(s_new,1);
B2=ones(s_new,1);

%Matrix computation for solving least squares problem
temp=(A'*A+eta1*eye(size(A,2)) + eta2*eye(size(A,2)));

%Perform iterations
 for iteration=1:maxit

     %Least square
     X=sp1(A,W,Z,y_vec,B1,B2,eta1,eta2,temp);

     %Nuclear norm minimization
      W=X+B1;
     [W,s_thld]=nuc_norm(W,s_x,s_new,lambda1,eta1);

     %L1 minimization
     Z = sp3(X,Z,D,B2,gamma,c);

     %Bregman variable update
     B1 = B1+X-W;
     B2 = B2+X-Z;

     %Computc O.F.
     obj_func(iteration)=0.5*norm(y_vec-A*X)+ lambda1*sum(s_
thld) + lambda2*sum(abs(X));

     %Convergence Criteria
     if iteration>10 && abs(obj_func(iteration)-obj_
func(iteration-1))<1e-8
         break
     end
 end

 %Recovered Matrix
 x_out = (reshape(X,s_x(1),s_x(2)));

 %Plot convergence behaviour
 plot(obj_func);
 title('Convergence Behaviour');
 xlabel('No. of iterations');
 ylabel('O.F value');

end
```

8

Combined Sparse and Low-Rank Recovery

In previous chapters, we learnt about sparse reconstruction and low-rank recovery techniques. In this concluding chapter of algorithms part, we will learn a few combined models involving both sparse and low-rank recovery. These do not always enjoy the theoretical constructs of sparse reconstruction and low-rank reconstruction but are practical powerful tools that yield very good results.

First, let us try to identify a few cases where such problems may arise. We will introduce a few of them now and will go deeper into them in the applications part of the book.

Consider the problem of video background-foreground separation. There are many computer vision-based techniques to address these problems, but we can also do it by using a neat signal processing formulation. In any video, the background is almost static—there is hardly any change. Also, the background fills the major part of the screen. The foreground occupies a smaller portion of the screen but is dynamic. Therefore, if we want to separate the background and the foreground from the entire video sequence, we can assume the background to be low-rank and the foreground to be sparse; that is, if the frames are stacked one after another in the video sequence, the background portion of the frames that do not change much with time will be linearly dependent and hence form a low-rank matrix. The foreground is small and hence can be assumed to be sparse with respect to the entire scene.

In this case, we have assumed that the foreground is sparse in itself; this is a simplifying assumption and we can do better. The foreground, even if not sparse, can be modeled as a sparse set of transform coefficients. This is because it is correlated in time domain, and hence, any transforms such as a Fourier or temporal differencing would produce a sparse set of coefficients for the foreground.

Now, imagine the problem of medical imaging. The data is acquired in a transform domain. But the usual model of foreground and background separation still holds. The background—the larger portion under field of view—is mostly static. The foreground is changing either due to motion or due to change in concentration (of some die). In such a case, we have to simultaneously reconstruct and separate the foreground and the background. Of course, it can be done in two phases: reconstruction followed by foreground-background separation, but compressed sensing (CS) allows us a single framework to solve the problem in one go.

The aforesaid problems pertain to the low-rank + sparse model. There is another model where the signal is both sparse and low-rank simultaneously. This arises mainly in dynamic magnetic resonance imaging (MRI). Dynamic MRI sequences can be modeled as sparse coefficients by optimally exploiting their spatial and temporal redundancies (for reconstruction via CS) or can be modeled as a low-rank matrix owing to the temporal redundancy (and implicit spatial correlation). The low-rank model by itself is more of a proof of concept and does not yield results better than sparse recovery. However, a combined low-rank and sparse model exhibits excellent results.

In the previous chapter, we studied about joint-sparse/row-sparse recovery of multiple measurement vectors. Such row-sparse problems are inherently low-rank (for matrices of reasonable size). This is because most of the rows are zeroes; hence, the rank of the matrix can never be more than the number of non-zero rows. There are studies showing that explicit modeling of the low-rank matrix for joint-sparse recovery problems actually helps.

Compressive Principal Component Pursuit

The simplest model for sparse + low-rank components is

$$X = L + S \tag{8.1}$$

where X is the signal, L is the low-rank component, and S is the sparse component. This model bears resemblance to principal component analysis (PCA) (the low-rank component); in such a scenario, the sparse portion is treated as outlier, making the PCA robust. Hence, (8.1) is also called robust PCA. The technique to solve this is called principal component pursuit.

The first generalization is to assume that the sparse component is not sparse in itself (such as an outlier) but has a sparse representation in the transform domain; this arises when the sparse component models foreground in a video separation problem. The second generalization of (8.1) is to assume that the data has been captured in some transform domain (e.g., Fourier for MRI). The entire data acquisition model can hence be expressed as,

$$y = \Phi X = \Phi(L + S) \tag{8.2}$$

Here, Φ is the measurement operator.

The solution for (8.2) is called compressive principal component pursuit.

Derivation of Principal Component Pursuit

In (8.2), we assumed a system that is not noisy; usually, there is some normal noise in the system. Therefore, we need to solve the following,

$$\min_{L,S} \|y-\Phi(L+S)\|_2^2+\lambda(\|L\|_*+\gamma\|\Psi(S)\|_1) \tag{8.3}$$

We employ the majorization-minimization (MM) technique to solve this; applying this on the cost function $\|y-\Phi(L+S)\|_2^2$ leads to the Landweber iterations,

$$\hat{L}+\hat{S}=L^k+S^k+\frac{1}{a}\Phi^T(y-\Phi(L^k+S^k)) \tag{8.4}$$

where k denotes the iteration number and a>*max eigenvalue* $(\Phi^T\Phi)$.

The Landweber iterations allow the cost function for the *kth* iteration to be written in the form,

$$\min_{L+S} \|L^k+S^k-L+S\|_2^2 \tag{8.5}$$

Therefore, using the Landweber iterations, we can express (8.3) in each iteration in the following form,

$$\min_{L,S} \|\hat{L}+\hat{S}-L+S\|_2^2+\frac{\lambda}{a}\left(\|L\|_*+\gamma\|\Psi(S)\|_1\right) \tag{8.6}$$

Since the sparse (S) and the low-rank (L) components are separable, the method of alternating directions can be used to decompose (8.6) into the following two problems,

$$L^{k+1}=\min_{L} \left\|\hat{L}+\hat{S}-L-S^k\right\|_2^2+\frac{\lambda}{a}\|L\|_* \tag{8.7}$$

$$S^{k+1}=\min_{S}\left\|\hat{L}+\hat{S}-L^{k+1}-S\right\|_2^2+\frac{\lambda\gamma}{a}\|\Psi(S)\|_1 \tag{8.8}$$

Now, (8.7) is minimized by singular value shrinkage,

$$L^{k+1}=U\ Soft_{\lambda/2a}(\Sigma)\ V^T,\ \text{where } \hat{L}+\hat{S}-S^k=U\Sigma V^T \tag{8.9}$$

$Soft_{\lambda/2a}(\Sigma)$ denotes the singular values after soft thresholding, that is, $Soft_{\lambda/2a}(\Sigma)=diag(\Sigma)\cdot\max(0,diag(\Sigma)-\lambda/a)$.

Solving (8.8) is exactly the same as solving a CS algorithm with analysis prior. It can be minimized in two steps:

Step 1: $z^{k+1} = (cz^k + \Psi(\hat{s}^k - \Psi^T z^k)) \cdot / \left(\frac{2a}{\lambda\gamma} \left| \Psi \hat{s}^k \right| + c \right)$
where $\hat{s}^k = vec(\hat{L} + \hat{S} - L^{k+1})$, $c > max\ eigenvalue(\Psi^T\Psi)$, and "." denotes element-wise operation.
Step 2: $s^{k+1} = \hat{s}^k - \Psi^T z^{k+1}, S^{k+1} = matrix(s^{k+1})$.

In a crisp fashion, the algorithm for solving the unconstrained problem can be described as follows:

ALGORITHM 1 SOLVING THE UNCONSTRAINED PROBLEM

Define: $J_k = \left\| y - \Phi(L^k + S^k) \right\|_2^2 + \lambda \left(\left\| L^k \right\|_* + \gamma \left\| \Psi(S^k) \right\|_1 \right)$
Initialize: S^0, L^0, and z^0.
While $\frac{J_k - J_{k+1}}{J_k + J_{k+1}} \geq Tol$

1. Landweber iteration: $\hat{L} + \hat{S} = L^k + S^k + \frac{1}{a}\Phi^T(y - \Phi(L^k + S^k))$
2. Update low-rank component: $L^{k+1} = U\ Soft_{\lambda/2a}(\Sigma)V^T$, where $\hat{L} + \hat{S} - S^k = U\Sigma V^T$
3. Update auxiliary variable: $z^{k+1} = (cz^k + \Psi(\hat{s}^k - \Psi^T z^k)) \cdot / (\frac{2a}{\lambda\gamma} \left| \Psi\ \hat{s}^k \right| + c)$
4. Update sparse component in vector form: $s^{k+1} = \hat{s}^k - \Psi^T z^{k+1}$
5. Form the sparse matrix: $S^{k+1} = matrix(s^{k+1})$

End

Sparse and Low-Rank Model

The robust PCA problem enjoys certain theoretical recovery guarantees, but the model we are going discuss now does not. The inverse problem that we intend to solve is as follows,

$$y = Ax + n \tag{8.10}$$

where the solution x is both sparse and low-rank. Depending on synthesis and analysis sparsity, we need to solve two different problems.

$$Synthesis: \|y - Ax\|_2^2 + \lambda_1 \|x\|_1 + \lambda_2 \|X\|_* \tag{8.11}$$

$$Analysis: \|y - Ax\|_F^2 + \lambda_1 \|Hx\|_p^p + \lambda_2 \|X\|_* \tag{8.12}$$

The symbols have their usual meanings.

To solve this problem, the first step is to replace the nuclear norm by its equivalent Ky-Fan norm, that is,

$$\|X\|_* = Tr(X^T X)^{1/2}.$$

Solving Synthesis Prior

For the synthesis prior problem, the function to be minimized is,

$$\min_x \|y - Ax\|_2^2 + \lambda_1 \|x\|_1 + \lambda_2 Tr(X^T X)^{1/2} \tag{8.13}$$

There is no closed-form solution to $J(x)$; it must be solved iteratively via MM. This leads to,

$$G_k'(x) = \|b - x\|_2^2 + \frac{\lambda_1}{a} \|x\|_1 + \frac{\lambda_2}{a} Tr(X^T X)^{1/2} \tag{8.14}$$

where $b = x_k + 1/a\, A^T(y - Ax_k)$ and a is the maximum eigenvalue of $A^T A$.

To minimize (8.14), we take its derivative,

$$2x - 2b + \frac{\lambda_1}{a} |x|^{-1} \cdot x + \frac{\lambda_2}{2a} (XX^T)^{-\frac{1}{2}} X \tag{8.15}$$

where "." denotes element-wise product.

Setting the gradient to zero, one gets,

$$(I + D)x = b \tag{8.16}$$

where $D = \lambda_1/2a\, Diag(|x|^{-1}) + \lambda_2/4a\, I \otimes (XX^T)^{-1/2}$.

Here, the *Diag* operator creates a diagonal matrix out of the vector $|x|^{-1}$.

The problem (8.16) represents a system of linear equations. It should be noted that the system $(I + D)$ is symmetric. Hence, it can be efficiently solved.

Based on this derivation, we propose the following algorithm to solve (8.16).

Initialize: $x_0 = 0$
Repeat until: $\|y - Ax\|_2^2 \le \varepsilon$
Step 1. $b = x_k + \frac{1}{a} A^T (y - Ax_k)$
Step 2. $D = \frac{\lambda_1}{2a} Diag(|x_{k-1}|^{-1}) + \frac{\lambda_2}{4a} I \otimes (X_{k-1} X_{k-1}^T)^{-1/2}$
Step 3. Compute x by solving $(I + D)x = b$
End

ALTERNATE DERIVATION BASED ON FOCUSS

In the synthesis prior problem, one can employ the FOCally Underdetermined System Solver (FOCUSS) approach to solve the constrained problem.

$$\min_X \|x\|_1 + \eta \|X\|_* \text{ subject to } y = Ax,\ x = vec(X) \tag{8.17}$$

As before, the nuclear norm is replaced by the equivalent Ky-Fan norm in the expression for its unconstrained Lagrangian form,

$$L(x, \lambda) = \|x\|_1 + \eta Tr(X^T X)^{1/2} + \lambda^T (y - Ax) \tag{8.18}$$

where λ is the vector of Lagrangian multipliers.

The Karush Kuhn Tucker (KKT) conditions for (8.18) are,

$$\nabla_x L(X, \lambda) = |x|^{-1} \cdot x + \eta q (XX^T)^{-\frac{1}{2}} X + A^T \lambda = 0 \tag{8.19}$$

$$\nabla_\lambda L(X, \lambda) = Ax - y = 0 \tag{8.20}$$

Now, (8.19) can be expressed as,

$$Wx + A^T \lambda = 0, \text{ where } W = pDiag(|x|^{-1}) + \eta q I \otimes (XX^T)^{-q/2} \tag{8.21}$$

Solving for x: $x = -W^{-1}A^T\lambda$

W is a block diagonal matrix with positive semi-definite blocks along the diagonal. Since W is positive semi-definite, the solution is not numerically stable. Such a problem was encountered while using FOCUSS for sparse signal recovery in CS; in order to reach a stable solution, W must be positive definite and hence must be regularized. Following these studies, we regularize W by adding a small term along the diagonal, that is, we replace,

$$W \rightarrow W + \varepsilon I \tag{8.22}$$

Here, ε is a small constant that regularizes the solution. This regularization also guarantees W' (and hence its inverse) to be positive definite. As $\varepsilon \rightarrow 0$, one arrives at the desired solution.

Solving for λ,

$$\lambda = -(AW^{-1}A^T)^{-1}y \tag{8.23}$$

Substituting, the value of λ, we get,

$$x = W^{-1}A^T(AW^{-1}A^T)^{-1}y \tag{8.24}$$

In order to efficiently compute x in each iteration, we re-write (8.24) as,

$$x = R\tilde{x}, \text{ where } \tilde{x} = (AR)^T((AR)(AR)^T)^{-1}y$$

Here, R is the Cholesky decomposition of W^{-1}. The decomposition exists since W^{-1} is a positive definite matrix. The reason we expressed (8.24) in the current form is because $\tilde{x}$ can be solved very efficiently by using the conjugate gradient (CG). Based on this modification, one has the following efficient algorithm to solve the Schatten-p norm minimization problem.

Initialize: $x_0 = A^T(AA^T)^{-1}y$, which is a least squares solution; define ε.

Repeat until stopping criterion is met:

Compute: $W_k = pDiag(|x_{k-1}|^{-1}) + \eta q I \otimes (X_{k-1}X_{k-1}^T)^{-1/2} + \varepsilon I$ and $R_k R_k^T = W_k^{-1}$.

Update: $\tilde{x}_k = (AR_k)^T((AR_k)(AR_k)^T)^{-1}y$ and $x_k = R\tilde{x}_k$. Reshape x_k to matrix form X_k.

Decrease: $\varepsilon = \varepsilon/10$ iff $\|x_k - x_{k-1}\|_2 \le tol$.

Solving Analysis Prior

The task is to solve (8.25). We express it in the following form:

$$\min_x \|y - Ax\|_F^2 + \lambda_1 \|Hx\|_1 + \lambda_2 Tr(X^T X)^{1/2} \tag{8.25}$$

Following the MM approach at each iteration, we need to solve,

$$\min_x \|B - X\|_2^2 + \frac{\lambda_1}{a}\|Hx\|_1 + \frac{\lambda_2}{a} Tr(X^T X)^{1/2} \tag{8.26}$$

where $B = X_k + 1/a\, A^T(Y - AX_k)$.

To minimize (8.26), we take its derivative,

$$2X - 2B + \frac{\lambda_1}{a} H^T \Omega H x + \frac{\lambda_2}{2a}(XX^T)^{-\frac{1}{2}} X, \text{ where } \Omega = diag(|Hx|^{-1}) \tag{8.27}$$

Setting the gradient to zero, one gets,

$$\left(I + \frac{\lambda_1}{2a} H^T \Omega H + \frac{\lambda_2}{4a} I \otimes (XX^T)^{-\frac{1}{2}}\right) x = b, \text{ where } b = vec(B) \tag{8.28}$$

Since (8.28) is not separable like the synthesis prior, finding the solution for the analysis prior problem is slightly more involved than the synthesis prior one. We express (8.28) as follows,

$$\left(M + \frac{\lambda_1}{2a} H^T \Omega H\right) x = b, \text{ where } M = I + \frac{\lambda_2}{4a} I \otimes (XX^T)^{-\frac{1}{2}} \tag{8.29}$$

Using the matrix inversion lemma,

$$\left(M + \frac{\lambda_1}{2a} H^T \Omega H\right)^{-1} = M^{-1} - M^{-1} H^T \left(\frac{2a}{\lambda_1}\Omega^{-1} + HM^{-1}H^T\right)^{-1} HM^{-1}$$

Therefore, we have the following identity,

$$x = M^{-1}b - M^{-1}H^T \left(\frac{2a}{\lambda_1}\Omega^{-1} + HM^{-1}H^T\right)^{-1} HM^{-1}b \tag{8.30}$$

Or equivalently,

$$z = \left(\frac{2a}{\lambda_1} \Omega^{-1} + HM^{-1}H^T \right)^{-1} HM^{-1}b \tag{8.31}$$

$$x = M^{-1}b - M^{-1}H^T z \tag{8.32}$$

Solving z requires solving the following,

$$\tilde{z} = \left(\frac{2a}{\lambda_1} \Omega^{-1} + HM^{-1}H^T \right)^{-1} HM^{-1/2}b, z = M^{-1/2}\tilde{z} \tag{8.33}$$

Here, $M^{-1/2}$ is the Cholesky decomposition of M. The decomposition exists since M is symmetric positive definite (follows from the definition of M in (8.29)).

It is possible to solve $\tilde{z}$ by the CG method [24]. Once $\tilde{z}$ is solved, finding the value of x is straightforward.

This derivation leads to the following iterative algorithm:

Initialize: $x_0 = 0$
Repeat until: $\|y - Ax\|_2^2 \le \varepsilon$
Step 1. $B = X_k + \frac{1}{a} A^T (Y - AX_k)$
Step 2. $M = I + \frac{\lambda_2}{4a} qI \otimes (XX^T)^{-1/2}$, $b = vec(B)$
Step 3. Solve: $\tilde{z} = (\frac{2a}{\lambda_1} \Omega^{-1} + HM^{-1}H^T)^{-1} HM^{-1/2}b$
Step 4. Compute: $z = M^{-1/2}\tilde{z}$
Step 5. Compute: $x = M^{-1}b - M^{-1}H^T z$
End

Suggested Reading

J. Wright, A. Ganesh, K. Min and Y. Ma, Compressive principal component pursuit, *Information and Inference*, 2 (1), 32–68, 2013.

S. G. Lingala, Y. Hu, E. V. R. DiBella and M. Jacob, Accelerated dynamic MRI exploiting sparsity and low-rank structure: k-t SLR, *IEEE Transactions on Medical Imaging*, 30 (5), 1042–1054, 2011.

A. Majumdar, Improved dynamic MRI reconstruction by exploiting sparsity and rank-deficiency, *Magnetic Resonance Imaging*, 31 (5), 789–795, 2013. (I.F. 2.0).

A. Majumdar, R. K. Ward and T. Aboulnasr, Non-convex algorithm for sparse and low-rank recovery: Application to dynamic MRI reconstruction, *Magnetic Resonance Imaging*, 31 (3), 448–455, 2013. (I.F. 2.0).

Appendix: MATLAB® Codes

Generalized Principal Component Pursuit

```
function [S, L] = L1NN(y, F, W, sizeImage, beta, err,
alpha, c)

% algorithm for solving the following optimization problem
% min nuclear_norm(L) + beta*||W(S)||_1
% subject to ||y-F(S+L)|_2 < err
% This is a generalized version of Pricipal Component
  Pursuit (PCP) where
% the sparsity is assumed in a transform domain and not in
  measurement
% domain. Moreover the samples obtained are lower
  dimensional projections.

% Inputs
% y - observation (lower dimensional projections)
% F - projection from signal domain to observation domain
% W - transform where the signal is sparse
% sizeImage - size of matrix or image
% beta - term balancing sparsity and rank deficiency
% err - related to noise variance
% alpha - maximum eigenvalue of F'F
% c - maximum eigenvalue of W'W
% Outputs
% S - sparse component
% L - low rank component

decfac = 0.5;
tol = 1e-4;
insweep = 50;
p = 0.5; q = 0.5;

if nargin < 6
    err = 1e-6;
end
if nargin < 7
    alpha = 1;
end
if nargin < 8
    c = 1;
end
```

```
L = zeros(sizeImage); % Low rank component
S = zeros(sizeImage); % Sparse component
sl = zeros(prod(sizeImage),1);
z = W(S(:),1);

lambdaInit = decfac*max(abs(F(y,2))); lambda = lambdaInit;

Sigma = svd(L);
f_current = norm(y-
F(S(:)+L(:),1)) + lambda*beta*norm(W(S(:),1),1) +
lambda*norm(diag(Sigma),1);

lambdaInit = decfac*max(abs(F(y,2))); lambda = lambdaInit;

while lambda > lambdaInit*tol
    for i = 1:insweep
        f_previous = f_current;

        % Landweber update
        sl = sl + (1/alpha)*F(y-F(S(:)+L(:),1),2);

        % Updating sparse component
        stilde = sl - L(:);
        z = (c*z + W(stilde-
W(z,2),1))./(((2*alpha)/(beta*lambda))*abs(W(stilde,1))+c);
% L1 Minimization
%          z = (c*z + W(stilde-
W(z,2),1))./(((2*alpha)/(beta*lambda))*abs(W(sti
lde,1)).^(2-q)+c);
        s = stilde - W(z,2);
        S = reshape(s,sizeImage);

        % Updating low rank component
        ltilde = sl-S(:);
        [U,Sigma,V] = svd(reshape(ltilde,sizeImage));
        Sigma = sign(diag(Sigma)).*max(0,abs(diag(Sigma))-
lambda/(2*alpha)); % Nuclear Norm Minimization
%          Sigma = sign(diag(Sigma)).*max(0,abs(diag(Sigma))-
abs(diag(Sigma).^(p-1)).*lambda/(2*alpha));
% Schatten p-norm Minimization
        L = U*diag(Sigma)*V';

        f_current = norm(y-F(S(:)+L(:),1)) +
lambda*beta*norm(W(S(:),1),1) + lambda*norm(Sigma,1);
```

```
        if norm(f_current-f_previous)/norm(f_current +
        f_previous)<tol
            break;
        end
    end

    if norm(y-F(S(:)+L(:),1))<err
        break;
    end
    lambda = decfac*lambda;
end
```

9

Dictionary Learning

So far, we have studied the topics where we assumed that the sparsifying basis is known. For example, images are known to be sparse in wavelet, discrete cosine transform (DCT), or curvelet; speech has a sparse representation in short-time Fourier transform; and biomedical signals can be sparsely represented by wavelet or Gabor transform. Such fixed mathematical transforms are generic and, as we can see, applicable to a wide variety of problems. All such transforms are defined for representing signals following some abstract structures, for example, smooth functions, functions with finite number of discontinuities, and piecewise linear functions. Since many natural signals approximately follow such structures, these transforms are popular in compressed sensing.

However, researchers always believed that adaptively learnt representations will be better for a particular class of problems. This is intuitive—wavelets can be used to sparsify both natural images and biomedical signals. But if we have a transform that is learnt to represent only natural images and another transform to represent only electrocardiogram (ECG) signals, we may expect better representation in such learnt transforms.

This idea goes back to the late 90s, when researchers in vision wanted to learn filters that mimic early portions of the human visual system. From a large number of image patches, they learnt these filters. They look similar to mathematical transform, but they are not exactly the same. An example of such learnt filters for natural images is shown in Figure 9.1.

Figure 9.1 is not really something from Olhausen and Field or Lee and Seung; rather, it is a modern version from K-SVD. One can see that, from the image patches, it is possible to learn representations that are similar to fixed transforms such as wavelet and DCT. Such learnt transforms are not generalizable to other domains but are better at representing a given domain.

FIGURE 9.1
(a) Image patched used for learning. (b) Left to right: filters learnt using K-SVD, wavelet, and DCT.

Dictionary Learning

Dictionary learning is not a new topic. It was introduced in late 90s as an empirical tool to learn filters. The usual understanding of dictionary learning is shown in Figure 9.2. The dictionary (D) and the coefficients (Z) are learnt from the data (X) such that the learnt dictionary and the coefficients can synthesize the data. Mathematically, this is represented as,

$$X = DZ \tag{9.1}$$

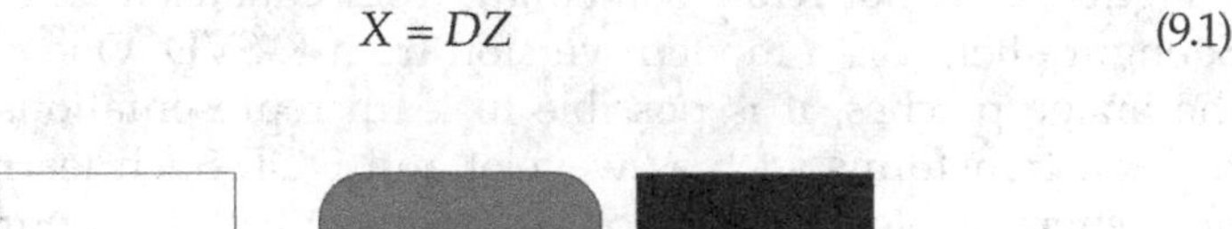

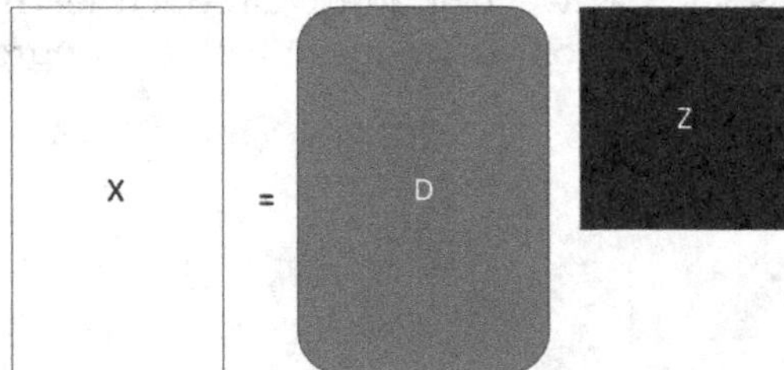

FIGURE 9.2
Schematic diagram for dictionary learning.

Early studies in dictionary learning focused on learning a basis for representation. There were no constraints on the dictionary atoms or on the loading coefficients. The method of optimal directions was used to learn the basis:

$$\min_{D,Z} \|X - DZ\|_F^2 \tag{9.2}$$

Here, X is the training data, D is the dictionary to be learnt, and Z consists of the loading coefficients.

It is easily solved by using alternate minimization. One starts by initializing the dictionary D. In each iteration (k), the coefficients are updated assuming the dictionary to be fixed (9.3), and then, the dictionary is updated assuming the coefficients to be fixed (9.4).

$$Z_k \leftarrow \min_{Z} \|X - D_{k-1}Z\|_F^2 \tag{9.3}$$

$$D_k \leftarrow \min_{D} \|X - DZ_k\|_F^2 \tag{9.4}$$

However, such alternate minimization can lead to degenerate solutions where either D is very large and Z is small, so that DZ equals X, or D is very small and Z is very large. To prevent such solutions, the atoms of the dictionary are normalized after every iteration. Alternately, one can also normalize the coefficients Z, but normalizing the dictionary is more prevalent.

For problems in sparse representation, the objective is to learn a basis that can represent the samples in a sparse fashion; that is, Z needs to be sparse. K-SVD is the most well-known technique for solving this problem. Fundamentally, it solves a problem of the form:

$$\min_{D,Z} \|X - DZ\|_F^2 \text{ such that } \|Z\|_0 \leq \tau \tag{9.5}$$

Here, we have abused the notation slightly; the l_0-norm is defined on the vectorized version of Z.

K-SVD proceeds in two stages. In the first stage, it learns the dictionary, and in the next stage, it uses the learned dictionary to sparsely represent the data. K-SVD employs the greedy (sub-optimal) orthogonal matching pursuit (OMP) [30] to solve the l_0-norm minimization problem approximately. In the dictionary learning stage, K-SVD proposes an efficient technique to estimate the atoms one at a time by using a rank-one update. The major disadvantage of K-SVD is that it is a relatively slow technique, owing to its requirement of computing the SVD (singular value decomposition) in every iteration.

There are other efficient optimization-based approaches for dictionary learning; these learn the full dictionary instead of updating the atoms separately. The usual formulation for such problems is as follows,

$$\min_{D,Z} \|X - DZ\|_2^2 + \lambda \|Z\|_1 \tag{9.6}$$

This too can be solved using alternating minimization.

$$Z_k \leftarrow \min_{Z} \|X - D_{k-1}Z\|_2^2 + \lambda \|Z\|_1 \tag{9.7}$$

$$D_k \leftarrow \min_{D} \|X - DZ_k\|_2^2 \tag{9.8}$$

As before, solving (9.8) is simple. It is a least square problem having a closed-form solution. The solution to (9.7), although not analytic, is well known in signal processing and machine learning literature. It can be solved using the iterative soft thresholding algorithm (ISTA). We already know how the algorithm for ISTA works.

Transform Learning

Dictionary learning is a synthesis formulation; that is, it learns a basis/dictionary along with the coefficients such that the data can be synthesized (Figure 9.3). There can be an alternate formulation where a basis is learnt to analyze the data to produce the coefficients. This is the topic of transform learning. The basic formulation is,

$$TX = Z \tag{9.9}$$

Here, T is the transform, X is the data, and Z is the corresponding coefficients. Relating transform learning to the dictionary learning formulation, we see that dictionary learning is an inverse problem, while transform learning is a forward problem.

One may be enticed to solve the transform learning problem by formulating,

$$\min_{T,Z} \|TX - Z\|_F^2 + \mu \|Z\|_0 \tag{9.10}$$

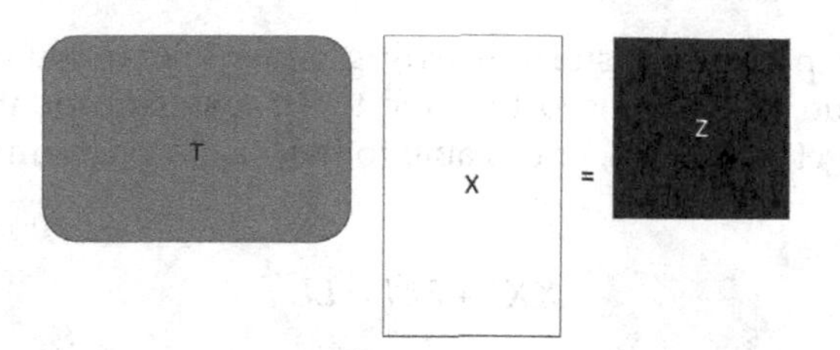

FIGURE 9.3
Schematic diagram for transform learning.

Unfortunately, such a formulation would lead to degenerate solutions; it is easy to verify the trivial solution $T = 0$ and $Z = 0$. In order to ameliorate this, one needs to ensure that the transform is never zero. This can be ensured from the following formulation,

$$\min_{T,Z} \|TX - Z\|_F^2 + \lambda\left(\varepsilon \|T\|_F^2 - \log\det T\right) + \mu \|Z\|_0 \quad (9.11)$$

The factor $-\log\det T$ imposes a full rank on the learned transform; this prevents the trivial solution. The additional penalty $\|T\|_F^2$ is to balance scale; without this, $\log\det T$ can keep on increasing, producing degenerate results in the other extreme.

As in dictionary learning, transform learning is solved using an alternating minimization approach.

$$Z \leftarrow \min_{Z} \|TX - Z\|_F^2 + \mu \|Z\|_0 \quad (9.12)$$

$$T \leftarrow \min_{T} \|TX - Z\|_F^2 + \lambda\left(\varepsilon \|T\|_F^2 - \log\det T\right) \quad (9.13)$$

Updating the coefficients, (9.12) is straightforward. It can be updated via one step of hard thresholding,

$$Z \leftarrow \left(abs(TX) \geq \mu\right) \odot TX \quad (9.14)$$

Here, $\odot$ indicates element-wise product.

For updating the transform, one can notice that the gradients for different terms in (9.13) are easy to compute. Ignoring the constants, this is given by,

$$\nabla \|TX - Z\|_F^2 = X^T (TX - Z)$$

$$\nabla \|T\|_F^2 = T$$

$$\nabla \log\det T = T^{-T}$$

In the initial paper on transform learning, a non-linear conjugate gradient-based technique was proposed to solve the transform update. Today, with some linear algebraic tricks, one is able to have a closed-form update for the transform.

$$XX^T + \lambda\varepsilon I = LL^T \tag{9.15}$$

$$L^{-1}YX^T = USV^T \tag{9.16}$$

$$T = 0.5R\left(S + (S^2 + 2\lambda I)^{1/2}\right)Q^T L^{-1} \tag{9.17}$$

The first step is to compute the Cholesky decomposition; the decomposition exists since $XX^T + \lambda\varepsilon I$ is symmetric positive definite. The next step is to compute the full SVD. The final step is the update step. One must notice that L^{-1} is easy to compute, since it is a lower triangular matrix.

Suggested Reading

B. A. Olhausen and D. J. Field, Sparse coding with an overcomplete basis set: A strategy employed by V1? *Vision Research*, 37 (23), 3311–3325, 1997.

D. D. Lee and H. S. Seung, Learning the parts of objects by non-negative matrix factorization, *Nature*, 401, 6755, 788–791, 1999.

K. Engan, S. Aase and J. Hakon-Husoy, Method of optimal directions for frame design, *IEEE International Conference on Acoustics*, Speech, and Signal Processing, 1999.

M. Aharon, M. Elad and A. Bruckstein, K-SVD: An algorithm for designing overcomplete dictionaries for sparse representation, *IEEE Transactions on Signal Processing*, 54 (11), 4311–4322, 2006.

S. Ravishankar and Y. Bresler, Learning sparsifying transforms, *IEEE Transactions on Signal Processing*, 61 (5), 1072–1086, 2013.

S. Ravishankar, B. Wen and Y. Bresler, Online sparsifying transform learning—Part I: Algorithms, *IEEE Journal of Selected Topics in Signal Processing*, 9 (4), 625–636, 2015.

S. Ravishankar and Y. Bresler, Online sparsifying transform learning—Part II: Convergence analysis, *IEEE Journal of Selected Topics in Signal Processing*, 9 (4), 637–746, 2015.

Appendix: MATLAB® Codes

Method of Optimal Directions

```
function [D, Z] = LearnDictionary(X, numOfAtoms)

Tol = 1e-3;
maxIter = 100;

if size(X,2) < numOfAtoms
    error('Number of atoms cannot be more than the number of
    samples');
end

% T = randperm(size(X,2));
% D = X(:,T(1:numOfAtoms)); % initialize dictionary with
randomly chosen samples
% for i = 1:numOfAtoms
%         D(:,i) = D(:,i)/norm(D(:,i));
% end

[Q,R]=qr(X);
D = Q(:,1:numOfAtoms);

% [U, S, V] = svd(X);
% D = S(:,1:numOfAtoms);

for iter = 1:maxIter
    % Feature Update
    Z = D \ X;
    % Z = [D; 0.1*eye(numOfAtoms)] \ [X; zeros
    (numOfAtoms,size(X,2))];

    Dprev = D;
    %Dictionary Update
    D = X / Z;
    for i = 1:numOfAtoms
        D(:,i) = D(:,i)/norm(D(:,i));
    end
    if norm(D-Dprev, 'fro') < Tol
        break
    end
end
```

Dictionary Learning

```
function [D, Z] = LearnSparseDictionary(X, numOfAtoms)

Tol = 1e-3;
maxIter = 100;

if size(X,2) < numOfAtoms
    display('Number of atoms cannot be more than the number
    of samples');
    exit
end

T = randperm(size(X,2));
D = X(:,T(1:numOfAtoms)); % initialize dictionary with
randomly chosen samples
for i = 1:numOfAtoms
        D(:,i) = D(:,i)/norm(D(:,i));
end

[Q,R]=qr(X);
D = Q(:,1:numOfAtoms);

% [U, S, V] = svd(X);
% D = S(:,1:numOfAtoms);

for iter = 1:maxIter
    % Feature Update
    Z = CooledDicIST(D, X);
    % Z = [D; 0.1*eye(numOfAtoms)] \ [X; zeros(numOfAtoms,
    size(X,2))];

    Dprev = D;
    %Dictionary Update
    D = X / Z;
    for i = 1:numOfAtoms
        D(:,i) = D(:,i)/norm(D(:,i));
    end
    if norm(D-Dprev, 'fro') < Tol
        break
    end
end
```

Sparse Coding (CooledDicIST)

```
function X = CooledDicIST(D, Y, err, insweep, tol, decfac)

% alpha = maxeig(H'H)
% for restricted fourier type of H, alpha > 1;


if nargin < 4
    err = 1e-8;
end
if nargin < 5
    insweep = 40;
end
if nargin < 6
    tol = 1e-3;
end
if nargin < 7
    decfac = 0.5;
end

alpha = max(svd(D)).^2;

sizeX = size(D'*Y);
X = zeros(sizeX);

lambdaInit = decfac*max(max(abs(D'*Y)));
lambda = lambdaInit;

f_current = norm(Y-D*X,'fro') + lambda*norm(X(:),1);

while lambda > lambdaInit*tol
     % lambda
    for ins = 1:insweep
        f_previous = f_current;

        B = X + (D'*(Y-D*X))/alpha;
        x = SoftTh(B(:), lambda/(2*alpha));
        X = reshape(x,sizeX);

        f_current = norm(Y-D*X,'fro') + lambda*norm(X(:),1);

        if norm(f_current-f_previous)/norm(f_current +
        f_previous)<tol
            break;
        end
    end
```

```
    if norm(Y-D*X,'fro')<err
        break;
    end
    lambda = decfac*lambda;
end

    function  z = SoftTh(s,thld)
        z = sign(s).*max(0,abs(s)-thld);
    end
end
```

Transform Learning

```
function [T, Z] = TransformLearning (X, numOfAtoms, mu,
lambda, eps)

% solves ||TX - Z||_Fro - mu*logdet(T) + eps*mu||T||_
Fro + lambda||Z||_1

% Inputs
% X          - Training Data
% numOfAtoms - dimensionaity after Transform
% mu         - regularizer for Tranform
% lambda     - regularizer for coefficient
% eps        - regularizer for Transform
% type       - 'soft' or 'hard' update: default is 'soft'
% Output
% T          - learnt Transform
% Z          - learnt sparse coefficients

if nargin < 5
    eps = 1;
end
if nargin < 4
    lambda = 0.1;
end
if nargin < 3
    mu = 0.1;
end

maxIter = 10;
type = 'soft'; % default 'soft'

rng(1); % repeatable
T = randn(numOfAtoms, size(X,1));
```

```
invL = (X*X' + mu*eps*eye(size(X,1)))^(-0.5);

for i = 1:maxIter

    % update Coefficient Z
    % sparse
    switch type
        case 'soft'
            Z = sign(T*X).*max(0,abs(T*X)-lambda); % soft
            thresholding
        case 'hard'
            Z = (abs(T*X) >= lambda) .* (T*X); % hard
            thresholding
    end
    % dense

    % Z = T*X;

    % update Transform T
    [U,S,V] = svd(invL*X*Z');
    D = [diag(diag(S) + (diag(S).^2 + 2*lambda).^0.5)
    zeros(numOfAtoms, size(X,1)-numOfAtoms)];
    T = 0.5*V*D*U'*invL;

end
```

10

Medical Imaging

Perhaps medical imaging has seen the largest application of compressed sensing (CS) since its inception. Application of sparse recovery-based techniques can make X-Ray computed tomography (CT) safer and magnetic resonance imaging (MRI) faster. There are a few applications of CS in other tomographic problems and in ultrasound imaging, but such studies are few in number. MRI and CT are by far the largest consumers of CS applications in medical imaging. In this chapter, we will study some well-known techniques in these areas. First, we will study about CT and then MRI.

X-Ray Computed Tomography

Figure 10.1 shows a simple CT imaging technique. The X-ray tube is the source; the subject is exposed to X-rays; these are attenuated by the subject, and the attenuated X-rays are received at the detector. Mathematically, the principle of CT is the following,

$$\int_L \mu^{(\bar{r})dl} = -\ln_{(I/I_0)} \tag{10.1}$$

where I_0 is the intensity of radiation at the source (known) and I is the intensity at the detector (measured). The distribution of μ, which is a material property, needs to be estimated; the limits of integration are also known.

Equation (10.1) holds for each projection. In CT, multiple projections are acquired from different angles. The problem is to estimate the underlying image (distribution of μ) from the detected measurements. We are not going into the mathematical details in this work; there are many introductory texts on this topic. From (10.1), we see that the relationship between μ and $\ln(I_0/I)$ for each projection is linear. Since each of the projections has a linear relationship between the measurement and the unknown, even without going

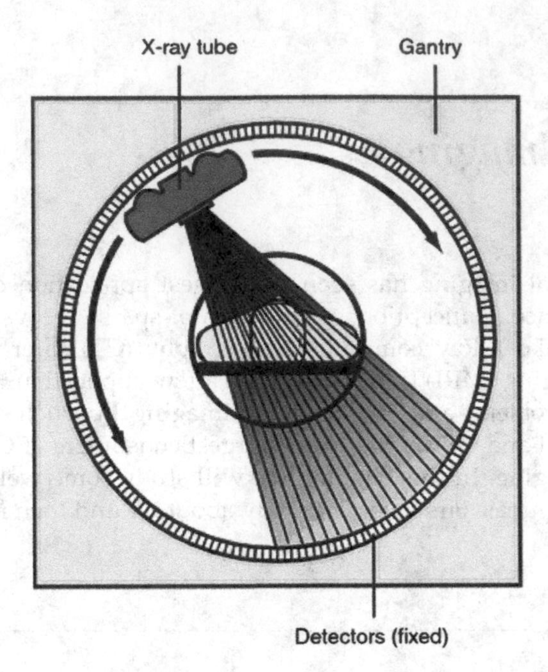

FIGURE 10.1
Simple X-Ray CT. http://medical-dictionary.thefreedictionary.com/computed+tomography.

into the detailed mathematics, we can infer that the CT imaging problem, being a collection of all the projections, will be a linear inverse problem:

$$f = Gx + \eta, \eta \sim N(0, \sigma^2) \tag{10.2}$$

where f is the sampled sinogram—this is what the CT scanner actually measures, G is the linear operator between the image and the sinogram, x is the image (distribution of μ), and η is the system noise.

The image formation process is not the focus of this chapter. The operator G is related to the Fourier transform via the Fourier slice theorem. For parallel beam X-rays, G is the well-known Radon transform. Unfortunately, in practice, CT scanners never use parallel beams; cone beams or fan beams are more common. In such situations, even though G is a linear operator, it is not the simple Radon transform. Modeling G as a fast linear operator is a topic in itself, but we are not going to delve into that. In this chapter, we will simply call G as the "X-ray transform."

Our interest is in the fact that the CT imaging is mathematically a linear inverse problem. In theory, there exists analytical solution for solving the inverse problem in the continuous domain; this is called the back propagation (BP). Unfortunately, the CT scanner can sample only discretely. Thus, the BP is not directly applicable; however, when there are infinitely many number of collected sinogram samples, a filtered back propagation (FBP) algorithm can be used to solve the reconstruction problem.

An analytical solution is theoretically elegant, but in this context, it demands infinitely many (or a very large number) of tomographic projections; more the number of tomographic projections, more is the X-ray dosage to which the subject is exposed. X-Ray CT device manufacturers claim that the radiation dosages from X-rays are within the US Food and Drug Administration (FDA)-approved limits. However, it is not the single scan that is hazardous. Most of the times, the ailing subject has to undergo scans multiple times; for example, if a cancer patient is diagnosed with a tumor (by a CT), he/she needs to check the progress of chemotherapy by undergoing further CT scans. The accumulated effect of multiple CT scans causes further complications.

It has been reported in the Archives of Internal Medicine that CT results in 30,000 additional cases of cancer every year and about 15,000 deaths from cancer in the USA. This figure excludes those cases where patients with former history of cancer were subjected to X-Ray CT. The *New England Journal of Medicine* published a study in 2007 that projected about 3 million excess cancer patients owing to X-Ray CT in the next 30 years. Children are more vulnerable to the risks of cancer from CT scans; a report published in the *American Journal of Roentgenology* in 2001 estimates that about 1500 children die from CT scans. In summary, X-Ray CT is not as safe as the medical imaging companies claim it to be—especially, when the subject suffers from an ailment that requires him/her to get CT scans regularly.

There needs to be a concerted effort in reducing X-Ray dosage in CT scans; in other words, the number of projections required for reconstructing the image needs to be reduced. However, when the number of CT projections is reduced, the reconstruction via FBP progressively suffers; the effect is shown in Figure 10.2 With the reduction in the number of projections, the inverse problem becomes less over-determined; when the number of projections is less than the image resolution, the problem becomes under-determined. In such scenarios, CS techniques become a viable approach for reconstruction. But before going into CS recovery techniques, we will review other techniques that preceded CS methods.

As seen before, CT reconstruction is a linear inverse problem (10.2). Given that the number of projections is typically large, the problem (10.2)

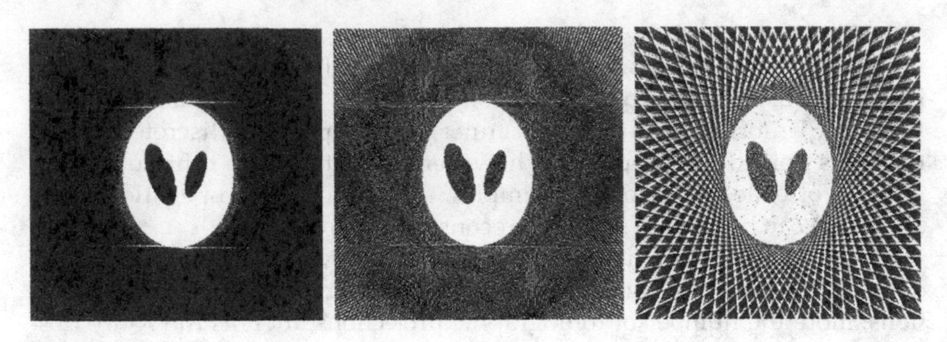

FIGURE 10.2
Left to right: very dense CT, dense CT, and parsimonious CT.

is over-determined; even with such large number of projections, FBP yields poor reconstruction. In such a case, solving the inverse problem by minimizing the least squares is a viable option:

$$\min_x \|f - Gx\|_2^2 \tag{10.3}$$

The Kaczmarz method for solving a system of over-determined equations was first introduced in the context of CT reconstruction (10.3). Kaczmarz method and other iterative techniques for solving the inverse problem (10.3) are active topics of research over the last two decades.

However, (10.3) also requires a large number of projections; albeit less the FBP. The question now arises: can we be even greedier and try to reduce the number of projections even further? CS gives an affirmative answers to this.

Compressed Sensing in Static Computed Tomographic Reconstruction

In CT, we are interested in reducing the number of projections required for reconstructing the image. The data acquisition model in CT is given by (10.2),

$$f = Gx + \eta$$

We are interested in reducing the number of projections to the extent that the linear system described above becomes under-determined. The underlying image happens to be sparse under a variety of linear transformations. CS exploits the transform domain sparsity of the underlying image in order to reconstruct it. One can assume that the image is sparse in a transform domain (orthogonal or tight-frame); incorporating this into (10.2) leads to,

$$f = G\Psi^T\alpha + \eta \quad (10.4)$$

where Ψ is the sparsifying transform and α is the vector of sparse transform coefficients. From this formulation, CS reconstruction is straightforward,

$$\min_{\alpha} \|\alpha\|_1 \text{ subject to } \|f - G\Psi^T\alpha\|_2^2 \leq \varepsilon \quad (10.5)$$

Instead of using the synthesis prior formulation (10.5), one can also solve the inverse problem by using analysis prior formulation. Usually, the total variation (TV) prior is used,

$$\min_{x} TV(x) \text{ subject to } \|f - Gx\|_2^2 \leq \varepsilon \quad (10.6)$$

Experimental Results

We show results on the Shepp-Logan phantom. For simplicity, we consider parallel beam CT, from which the sinogram can be easily simulated by using the Radon transform. We compare two methods:

1. Synthesis prior [10.5]
2. TV minimization [10.6]

To make it realistic, all images have been corrupted with 10% Gaussian white noise.

The reconstruction error is measured in terms of normalized mean squared error (NMSE). The reconstruction error is showed in the Table 10.1.

We find that TV minimization used slightly better results, but this may not be the case always. We have used the Shepp-Logan phantom, which has nicely defined contours and hence is particularly suitable for TV reconstruction (Figure 10.3).

TABLE 10.1

NMSE for Different Number of Projections

Method	30 lines	60 lines	90 lines
Synthesis prior	0.45	0.31	0.26
TV minimization	0.42	0.30	0.21

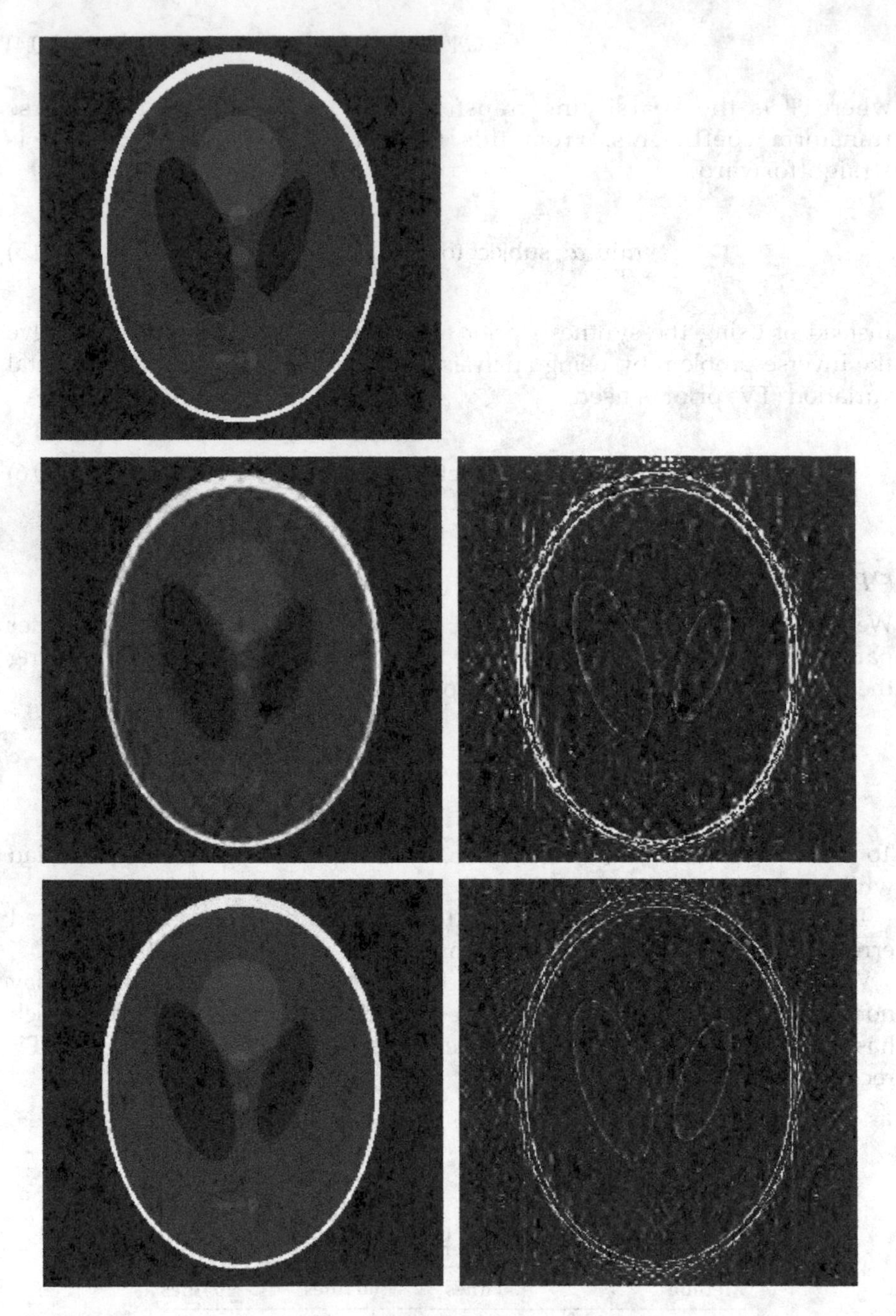

FIGURE 10.3
Top to bottom: ground-truth, synthesis prior, and TV minimization. Left: reconstructed image, and right: difference image.

Compressed Sensing in Dynamic Computed Tomography

For dynamic CT, the sinogram is sampled in an interleaved fashion, so the measurement matrix changes with time. The data acquisition model for the tth frame is as follows,

$$f_t = G_t x_t + \eta \tag{10.7}$$

We will study a popular technique for dynamic CT reconstruction. The first step is to generate a static FBP reference image (x_0) from the interleaved projections. Once this reference image is computed, the reconstruction of the tth frame is solved via the following optimization problem,

$$\min_x \alpha \left\| \Psi_1 (x_t - x_0) \right\|_p^p + (1-\alpha) \left\| \Psi_2 x_t \right\|_p^p \text{ subject to } \left\| f_t - G_t x_t \right\|_2^2 \leq \varepsilon \tag{10.8}$$

where Ψ_1 and Ψ_2 are sparsifying transforms (wavelet or gradient). The l_p-norm ($0 < p \leq 1$) is the sparsity-promoting objective function. There are two sparsity-promoting terms. The first term assumes that the difference between the current frame and the reference image is sparse in Ψ_1. The second term assumes that the tth frame is sparse in Ψ_2. The scalar α controls the relative importance of the two sparsity-promoting terms.

This formulation (10.8) is called prior image constrained compressed sensing (PICCS). This was originally developed with convex sparsity-promoting l_1-norm but was later shown to yield even better results with non-convex (NC) l_p-norm (NCPICCS). It should be noted that even though the frames are reconstructed separately, this is an offline technique because the reference image x_0 can be generated only after the full sequence has been collected.

Magnetic Resonance Imaging Reconstruction

MRI is a versatile medical imaging modality that can produce high-quality images and is safe when operated within approved limits. The main challenge that MRI faces today is its comparatively long data acquisition time. This poses a problem from different ends. For the patient, this is uncomfortable because he/she has to spend a long period of time in a claustrophobic environment (inside the bore of the scanner).[1] Thus, there is always the

[1] Hitachi has recently introduced the Patient Active Comfort Technology for scanning pediatric, geriatric, and claustrophobic subjects.

requirement of an attending technician to look after the patient's comfort. To make matters worse, the scanner is relatively noisy owing to the periodic switching on and off of the gradient coils.[2]

However, patient discomfort is not the only issue. As the data acquisition time is long, any patient movement inside the scanner results in unwanted motion artifacts in the final image; some of these movements are inadvertent, such as breathing.[3] Even these small movements may hamper the quality of images.

It is not surprising that reducing the data acquisition time in MRI has been the biggest challenge for the past two decades. The work is far from complete. Broadly speaking, there are two approaches to reduce the data acquisition/scan times: the hardware-based approach and the software-based approach. Initial attempts to reduce the scan time were hardware-based methods, where the design of the MRI scanner had to be changed in order to acquire faster scans. The multi-channel parallel MRI technique is the most well-known example of this exercise. Unfortunately, research and implementation of the hardware-based acceleration techniques were expensive and time-consuming; this is obvious, as new scanners had to be designed, built, and tested. Moreover, the collateral damages caused by such hardware-based acceleration techniques were not trivial. For example, multi-channel parallel MRI has been around for about 20 years. Reconstructing images from such scanners requires knowledge of sensitivity profiles of individual coils; this sensitivity information is never fully available. Therefore, reconstructing images from such parallel multi-channel MRI scans remains an active area of research even today.

Software-based acceleration methods do not require any modifications of the hardware of the scanner. All MRI scanners are controlled by a computer, which controls the pulse sequences and sampling and also reconstructs the images. Software-based acceleration only makes changes to pulse sequence control, sampling control, and the reconstruction modules to achieve faster scans. One well-known example of software-based acceleration is the introduction of gradient echo sequences against the earlier spin echo sequences. However, in order to accommodate faster gradient echo sequences (compared with the spin echo), one needs to trade-off signal-to-noise ratio (SNR).

Hardware-based acceleration techniques have reached their limits in the last decade. The key to future improvements is software-based acceleration. Standing on the solid theoretical background of CS, in the last decade, there has been a plethora of work in software-based acceleration. In the rest of the

[2] Recently, GE has introduced the Silent Scan technology; the noise in the scanner is around 77 dB, which is only slightly higher than the ambient 74 dB ambient noise.

[3] Siemens has introduced the new VIBES technology that enables various scans under free breathing.

section, we will learn how CS has revolutionized MRI reconstruction literature during this period. To begin with, we will concentrate on the problem of single-channel static MRI reconstruction. This is the most widely used MRI modality and the standard tool in clinical diagnosis. Technically speaking, this is the simplest MR imaging modality—the subject under study does not move, and only a single uniform channel is used to capture the image. Thus, the issue of sensitivity profile estimation does not arise. All other modifications, such as multi-channel parallel MRI and dynamic MRI, are extensions of this basic single-channel static MRI. Thus, understanding this modality is of fundamental importance to us.

Single-channel static MRI is further segregated into two parts: single-echo MRI and multi-echo MRI. Single-echo MRI is routinely used for medical diagnosis. Multi-echo MRI is a niche area and is not very common. It is applicable for quantitative MRI and some other problems. In the first part of this chapter, we will discuss the techniques for single-echo MRI reconstruction. This is because multi-echo MRI reconstruction is an extension of single-echo MRI, and thus, we need to learn the former first.

Single-Echo Static Magnetic Resonance Imaging Reconstruction

In MRI, the scanner samples the Fourier frequency space of the object under study. The sampling is discrete, and hence, the data acquisition model is expressed as follows,

$$y_{m\times 1} = F_{m\times n}x_{n\times 1} + \eta_{m\times 1}, \eta \sim N(0,\sigma^2) \tag{10.9}$$

where y is the collected Fourier frequency samples, x is the image (to be reconstructed), F is the Fourier mapping from the image space to the frequency space, and η is the noise that is assumed to be normally distributed.

The underlying object "x" may be an image or a volume but is represented as a vector in (10.9) by row or column concatenation. In MRI, for historical reasons, the Fourier frequency space is called the "K-space." In single-echo MRI reconstruction, the problem is to reconstruct the image (for ease, we will call it "image" for both two-dimensional [2D] images and three-dimensional [3D] volumes), given the K-space samples y. This is a classic linear inverse problem.

If the K-space sampling is performed uniformly on a regular Cartesian grid, the reconstruction is trivial, provided enough samples are collected, that is, when $n = m$. In this case, (10.1) is a determined system and reconstruction is easy—one just needs to apply the inverse fast Fourier transform (FFT) on the collected K-space samples in order to get the reconstructed image, that is,

$$\hat{x} = F^{-1}y \tag{10.10}$$

However, uniform Cartesian K-space sampling is time-consuming and is the main source of delay in MRI scans. Software-based acceleration techniques change the sampling grid in order to speed up the scans. Previously, non-Cartesian spiral, radial, and Rosetta sampling trajectories had been used to accelerate scans. The data acquisition speed is increased with such sampling trajectories, but the image reconstruction is not so simple anymore—one cannot use an inverse FFT to recover the image. Generally, there are two ways to solve (10.9). The first method is non-iterative. It interpolates the samples from the non-Cartesian grid (radial, spiral, and Rosetta) to a uniform Cartesian grid, from which the image is reconstructed via inverse FFT. But interpolation from the non-Cartesian to the Cartesian grid incurs errors. Thus, this simple method is not always recommended. The second method is to directly solve the inverse problem (10.9) as follows,

$$\hat{x} = \min_x \|y - Fx\|_2^2 \tag{10.11}$$

The solution to (10.11) is iterative. This is because, in practice, the Fourier operator is not available explicitly as a matrix, and hence, the left pseudoinverse cannot be employed to solve (10.3). Besides, there are also stability issues, since the non-Cartesian sampling trajectories make the Fourier mapping ill-conditioned. Hence, in practice, (10.11) is solved iteratively (and mostly with regularization). It must be remembered that for such non-Cartesian sampling trajectories, the number of K-space samples collected is always greater than the number of pixels/voxels in x, that is, $m > n$; thus, (10.11) is over-determined. This is true for both non-iterative gridding-based techniques and iterative least squares solutions (10.11). But even though non-Cartesian sampling trajectories require more number of samples (over-determined system) than Cartesian sampling (determined system), the time taken to complete the scan using non-Cartesian trajectories is usually less.

From the brief discussion, we have learnt so far that there is no free lunch; when the data acquisition time is small (non-Cartesian), the reconstruction is harder (gridding or iterative solution), but when the data acquisition time is larger (Cartesian), the reconstruction is trivial. In engineering problems, we always experience such trade-offs.

With the advent of CS, researchers in MRI reconstruction went one step ahead. CS showed that it is possible to reconstruct MR images from far fewer samples than were previously required—for both Cartesian and non-Cartesian sampling. By CS, it is possible to reconstruct the images fairly accurately, even when the number of K-space samples is smaller than the number of pixels in the image, that is, when $m < n$ in (10.1). As the scan time is dependent on the number of samples collected, this new discovery (CS reconstruction) meant that it was possible to reduce the scan time even further than was previously deemed possible.

The simplest way to solve the MRI reconstruction problem is to assume that the image is sparse in some transform domain, such as wavelet, and employ l_1 minimization to solve for the sparse transform coefficients,

$$\min_{\alpha} \|\alpha\|_1 \text{ subject to } \|y - FW^T\alpha\|_2^2 \leq \varepsilon \quad (10.12)$$

Here, $W = \Psi$ is the wavelet transform that sparsifies the MR image. Once the wavelet coefficients are solved, the image is reconstructed via the synthesis equation as follows,

$$x = W^T\alpha \quad (10.13)$$

MR images can be modeled in two different ways: (i) as smooth functions with finite number of curvilinear discontinuities, or (ii) as piecewise smooth functions. The first model leads to sparse representation of images in wavelet transform domain, while the second model leads to a sparse gradient. One can use information only about the sparse gradients while reconstructing MR images. But it is known that using the sparsity of the gradient information does not usually lead to good reconstruction results; sparsity-promoting l_1-norm over the wavelet coefficients is better than the TV norm (gradient sparsity).

As in the case of CT, MRI reconstruction can have two formulations. The optimization problem used for solving the synthesis prior formulation is given by (10.12). The TV norm is an analysis prior formulation,

$$\min_{\alpha} TV(x) \text{ subject to } \|y - Fx\|_2^2 \leq \varepsilon \quad (10.14)$$

Here, $TV(x) = \sum\left(|D_h x| + |D_v x|\right)$, and D_h and D_v are the horizontal and vertical differentiating operators, respectively. The first comprehensive study on this topic, dubbed as Sparse MRI, proposed combining the two priors—analysis (TV) and synthesis (wavelet); this leads to the following formulation,

$$\min_{\alpha} \|\alpha\|_1 + \gamma TV(x) \text{ subject to } \|y - Fx\|_2^2 \leq \varepsilon, \; x = W^T\alpha \quad (10.15)$$

Here, the objective function is a sum of l_1-norm and the TV norm. The l_1-norm promotes sparsity of the image in the wavelet domain, and the TV norm promotes gradient sparsity. The scalar γ controls the relative importance of the two terms; this needs to be specified by the user.

Multi-echo Magnetic Resonance Imaging

MRI is a versatile imaging modality. One can change the scan parameters, such that, for one set of parameters, the gray matter appears white, and for a different set of parameters, it can be made to appear black. In most cases, this versatility is good. But it also means that MR images are not quantitative. X-Ray CT is a quantitative imaging modality; there is one-to-one correspondence between tissues and the pixel/voxel values (Hounsfield numbers). This correspondence never changes. Such an imaging modality is "quantitative" where there is always a one-to-one correspondence between tissues and pixels/voxels.

Owing to its versatility, MRI has one-to-many correspondence and hence is not quantitative. The MR image is dependent on the scanning parameters.

In recent times, there has been an effort to produce quantitative MR images. It is not possible to obtain such images directly, unlike X-Ray CT. One needs to generate such images indirectly. The scan parameters are varied, and for each parametric value, an image is acquired. Once a set of such images has been acquired, multi-exponential curve fitting is done through each pixel/voxel across the parameter space. As a result of the curve fitting, one gets the tissue parameters. These maps of tissue parameters are the quantitative MR images. Such maps are quantitative because they directly show the tissue parameters and do not change (they have a one-to-one correspondence between tissues and points on the map).

The problem with generating such quantitative images is the prolonged scan time. Acquiring K-space samples for one image was bad enough, and now, one needs to collect K-space samples for multiple images/echoes. In this section, we will learn how the scan time can be reduced for the multi-echo MRI acquisition.

Multi-echo MRI is not only useful for the sake of quantitative imaging. During diagnosis, the doctor is interested in the tissue contrasts. Higher the contrast, easier it is for the doctor to diagnose. It is not possible to have good tissue contrasts between all tissues under the field of view (FoV). In general, it is only possible to maximize the contrast between two tissues; however, in some special cases, it is possible to maximize the contrast between three tissues at most. Thus, in order to maximize tissue contrast between multiple tissues in the FoV, multiple images/echoes must be gathered. This is another necessity for multi-echo MRI. To have a better understanding of the problem, we will briefly review the physics of MR image formation.

Physics of Magnetic Resonance Image Contrast

In MRI, the signal contrast is generated in a variety of ways. But the proton density (PD) weighting, T_1 (spin-lattice relaxation time) weighting, and T_2 (spin-spin relaxation time) are the most basic contrast-generating mechanisms. Apart from them, other factors such as magnetic susceptibility, tissue saturation, and diffusion can also be utilized for generating contrast.

However, in this chapter, we will concentrate on the basic ones: PD, T_1 weighting, and T_2 weighting.

The MR signal intensity of a particular tissue is governed by the following equation,

$$I = \rho_0 \left(1 - e^{-T_R/T_1}\right) e^{-T_E/T_2} \tag{10.16}$$

where ρ_0 is the PD (spin density), T_R is the repetition time, and T_E is the echo time of the applied magnetization. The T_R and the T_E are the scanning parameters.

The contrast between two tissues (A and B) in the MR image is the difference between the intensities,

$$C_{AB} = I_A - I_B = \rho_{0,A} \left(1 - e^{-T_R/T_{1,A}}\right) e^{-T_E/T_{2,A}} - \rho_{0,B} \left(1 - e^{-T_R/T_{1,B}}\right) e^{-T_R/T_{2,B}} \tag{10.17}$$

For PD weighting, the effects of T_1 and T_2 need to be nullified. This is achieved by making,

$$T_R \gg T_{1,A}, T_{1,B}, \Rightarrow e^{T_R/T_{1,A}}, \quad e^{T_R/T_{1,B}} \approx 0$$

$$T_E \ll T_{2,A}, T_{2,B}, \Rightarrow e^{T_E/T_{2,A}}, \quad e^{T_E/T_{2,B}} \approx 1$$

So, the contrast is,

$$C_{A,B} \approx \rho_{0,A} - \rho_{0,B} \tag{10.18}$$

It is not possible to obtain different amounts of PD weighting and thus is not of interest to us. We have included it for the sake of completion. However, PD is the fundamental governing term in MRI and all other types of MR signal acquisition include the effect of PD.

For T_1 weighting, the effect of T_2 weighting needs to be nullified (it is not possible to remove the effect of PD weighting). This is achieved by,

$$T_E \ll T_{2,A}, T_{2,B}, \Rightarrow e^{T_E/T_{2,A}}, e^{T_E/T_{2,B}} \approx 1$$

Therefore, the contrast between tissues A and B is,

$$C_{AB} = \rho_{0,A} \left(1 - e^{-T_R/T_{1,A}}\right) - \rho_{0,B} \left(1 - e^{-T_R/T_{1,B}}\right) \tag{10.19}$$

To maximize the contrast between the two tissues, (10.19) needs to be maximized with respect to T_R, and this value is,

$$T_{R,opt} = \frac{\ln\left(\frac{\rho_{0,B}}{T_{1,B}} - \frac{\rho_{0,A}}{T_{1,A}}\right)}{\left(\frac{1}{T_{1,B}} - \frac{1}{T_{1,A}}\right)} \tag{10.20}$$

It can be seen that the value of repetition time (T_R) that maximizes the contrast between two tissues is dependent on the tissue parameters (PDs and spin-lattice relaxation time). Therefore, in order to maximize the contrasts between different pairs of tissues in the FoV, different T_1-weighted images with different repetition times need to be taken. Such multi-echo T_1-weighted images will be useful for diagnostic MRI. In quantitative MRI, one is interested in the spin-lattice relaxation times of the tissues. It is possible to find this by fitting a multi-exponential curve through the bunch of acquired multi-echo images for each pixel/voxel. The resultant plot with spin-lattice relaxation times at each position is the T_1 map.

Similarly, one can have multi-echo T_2-weighted images. For T_2 weighting, the effect of T_1 weighting needs to be nullified. This is achieved by,

$$T_R \gg T_{1,A}, T_{1,B}, \Rightarrow e^{T_R/T_{1,A}}, e^{T_R/T_{1,B}} \approx 0$$

Thus, the contrast between two T_2-weighted tissues is,

$$C_{AB} = \rho_{0,A} e^{-T_E/T_{2,A}} - \rho_{0,B} e^{-T_E/T_{2,B}} \tag{10.21}$$

The optimal value of the echo time (T_E) that maximizes the contrast between the tissues is,

$$T_{E,opt} = \frac{\ln\left(\frac{\rho_{0,B}}{T_{2,B}} - \frac{\rho_{0,A}}{T_{2,A}}\right)}{\left(\frac{1}{T_{2,B}} - \frac{1}{T_{2,A}}\right)} \tag{10.22}$$

Like T_1-weighted images, one also needs to acquire T_2-weighted images. The purpose is the same. In order to visually analyze what it means by multi-echo MRI, T_2-weighted images of the transverse section of a human brain are shown in Figure 10.4.

The discussion in this section can be found in any textbook on MRI physics. We have repeated the discussion for the interest of a wider readership.

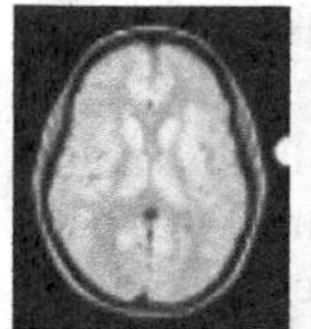 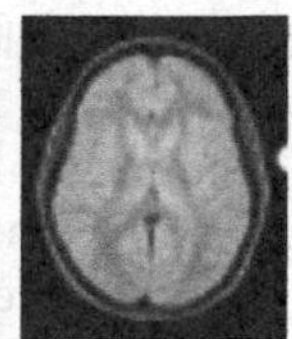 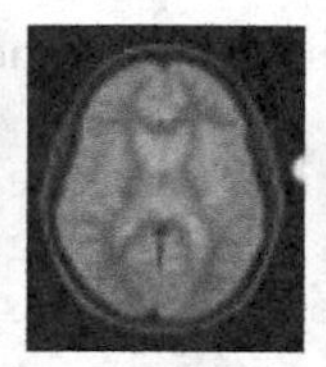 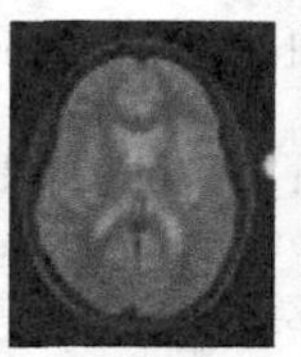 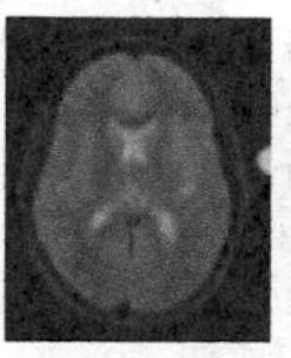 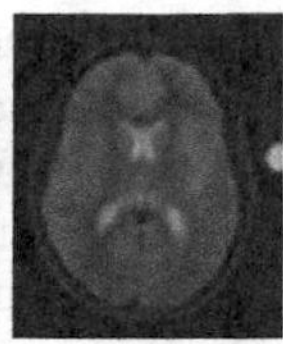

FIGURE 10.4
Images of same cross-section of brain with different T_2 weightings. (Courtesy Piotr Kozlowski.)

Group-Sparse Reconstruction of Multi-echo Magnetic Resonance Imaging

In this section, we will discuss everything with respect to T_2-weighted imaging. But the discussion is also pertinent to T_1-weighted MRI. In T_2-weighted MR imaging, multiple echoes of the same anatomical slice with varying echo times are acquired. The objective is to reconstruct the multi-echo T_2-weighted MR images. Generally, this kind of data is acquired for computing T_2 maps.

In conventional MR imaging, the echoes are acquired by fully sampling the K-space on a rectangular grid and applying 2D inverse FFT to reconstruct the image. Such conventional K-space sampling methods become prohibitively slow for practical multi-echo T_2-weighted MR imaging; for example, acquiring 32 echoes while imaging a single slice of the rat's spinal cord by using Carr-Purcell-Meiboom-Gill (CPMG) sequence takes about 40 minutes; therefore, imaging only 4 slices (without interleaving) takes around 3 hours.

For single 2D MR images, the scan time can be reduced by randomly acquiring only a subset of full K- space. The images from such partially sampled K-space data can be reconstructed by using CS-based techniques. CS-based techniques utilize spatial correlation (leading to sparsity of the image in a transform domain) in the MR image in order to reconstruct it.

Such CS-based technique can be directly applied to each of the echoes of the multiple T_2-weighted images. However, this is not the best possible approach. The multiple T_2-weighted images are correlated with each other. If the images are reconstructed piecemeal, this inter-echo correlation cannot be exploited during reconstruction. In order to get the maximum out of reconstruction, information about both intra-image spatial correlation and inter-image correlation needs to be utilized; instead of reconstructing each image individually, all the T_2-weighted images will be jointly reconstructed.

Contrast between various tissues in the MR image is dependent on the T_2 weighting; that is, for a particular T_2 weighting, the contrast between two tissues A and B may be high, while for another T_2 weighting, the contrast may be low. In multi-echo T_2-weighted imaging, each image varies from the other in their contrast between the tissues. Since all the T_2-weighted images actually correspond to the same cross-section, they are highly correlated among themselves.

As mentioned earlier, in T_2-weighted imaging, one acquires multiple images of the same cross-section by varying the T_2 weighting. Assume that the K-space data for N such weightings have been acquired. This is represented by,

$$y_i = F_i x_i + \eta_i, i = 1...N \tag{10.23}$$

where F_i represents that we can have a different sampling pattern for each T_2 weighting.

The problem is to reconstruct the T_2-weighted images (x_is), given their K-space samples (y_is).

A straightforward application of CS will repeatedly apply sparse recovery to reconstruct the T_2-weighted individually (for each i). As mentioned earlier, this is not optimal, since it does not account for inter-image correlation. Better reconstruction can be obtained when both intra-image spatial correlation and inter-image correlation in order to improve reconstruction accuracy.

Incorporating the wavelet transform, (10.23) can be expressed concisely as follows,

$$\tilde{y} = \Phi\tilde{\alpha} + \tilde{\eta} \tag{10.24}$$

where $\tilde{y} = \begin{bmatrix} y_1 \\ ... \\ y_T \end{bmatrix}, \Phi = \begin{bmatrix} FW^T & 0 & ... \\ 0 & ... & ... \\ 0 & ... & FW^T \end{bmatrix}, \tilde{\alpha} = \begin{bmatrix} \alpha_1 \\ ... \\ \alpha_T \end{bmatrix}$ and $\tilde{\eta} = \begin{bmatrix} \eta_1 \\ ... \\ \eta_T \end{bmatrix}$

The wavelet transform effectively encodes edges in an image. When the edge is more pronounced, the wavelet coefficients are high. When the edge is not so sharp, the wavelet coefficients are low. In smooth areas, the wavelet coefficients are zero. The following toy example illustrates the fact.

Consider a small image matrix with sharp boundary for a particular T_2 weighting (Figure 10.5a). The 0s correspond to tissue B, and the 1s correspond to tissue W. The wavelet transform (Figure 10.5b). Captures the vertical discontinuity between W and B perfectly. The values in the last column should be ignored; the wavelet transform assumes that the signal is periodic and computes the boundary between W and B.

Figure 10.5 shows that when the tissue boundary is pronounced, the wavelet transform along the boundary is high. Now, consider a different value of T_2 weighting, such that the contrast between W and B is less pronounced (Figure 10.6a). The wavelet transform of this matrix is shown in Figure 10.6b. We can see that even though the position of the high-valued transform coefficients is the same, their values have changed. As the contrast has reduced, the value of the wavelet coefficients has also reduced.

(a)								(b)							
1	1	1	1	0	0	0	0	0	0	0	1	0	0	0	-1
1	1	1	1	0	0	0	0	0	0	0	1	0	0	0	-1
1	1	1	1	0	0	0	0	0	0	0	1	0	0	0	-1
1	1	1	1	0	0	0	0	0	0	0	1	0	0	0	-1
1	1	1	1	0	0	0	0	0	0	0	1	0	0	0	-1
1	1	1	1	0	0	0	0	0	0	0	1	0	0	0	-1
1	1	1	1	0	0	0	0	0	0	0	1	0	0	0	-1
1	1	1	1	0	0	0	0	0	0	0	1	0	0	0	-1

FIGURE 10.5
First T_2 weighting: 2a. Tissue boundary: 2b. Wavelet coefficients.

(a)								(b)							
0.75	0.75	0.75	0.75	0.25	0.25	0.25	0.25	0	0	0	0.5	0	0	0	-0.5
0.75	0.75	0.75	0.75	0.25	0.25	0.25	0.25	0	0	0	0.5	0	0	0	-0.5
0.75	0.75	0.75	0.75	0.25	0.25	0.25	0.25	0	0	0	0.5	0	0	0	-0.5
0.75	0.75	0.75	0.75	0.25	0.25	0.25	0.25	0	0	0	0.5	0	0	0	-0.5
0.75	0.75	0.75	0.75	0.25	0.25	0.25	0.25	0	0	0	0.5	0	0	0	-0.5
0.75	0.75	0.75	0.75	0.25	0.25	0.25	0.25	0	0	0	0.5	0	0	0	-0.5
0.75	0.75	0.75	0.75	0.25	0.25	0.25	0.25	0	0	0	0.5	0	0	0	-0.5
0.75	0.75	0.75	0.75	0.25	0.25	0.25	0.25	0	0	0	0.5	0	0	0	-0.5

FIGURE 10.6
Second T_2 weighting: 2a. Tissue boundary: 2b. Wavelet coefficients.

This example is to corroborate the fact that as long as the anatomy of the brain slice does not change, the positions of the high-valued wavelet transform coefficients do not change for different T_2 weightings. Mathematically, this means that there should be a high degree of mutual correlation between the wavelet transforms of any two T_2-weighted images of the same anatomical cross-section. Figure 10.7 shows the scatter plot between the wavelet coefficients of two randomly chosen T_2-weighted images of rat's spinal cord. The plot shows that the correlation is almost perfectly linear.

The linear relationship corroborates our physical understanding of the fact that wavelet coefficients of two T_2-weighted images have similar valued coefficients at similar positions.

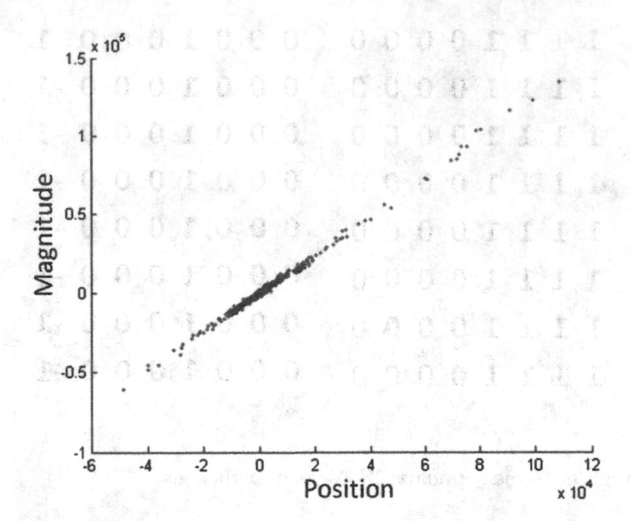

FIGURE 10.7
Scatter plot of wavelet coefficients of two T_2-weighted images of rat's spinal cord.

Group-Sparse Synthesis Prior

The wavelet coefficient vector for each image (αi) corresponding to different T_2 weightings is of length n (assuming for the time being that the wavelet transform we are considering is orthogonal). The combined wavelet coefficients α can be grouped according to their positions, as shown in Figure 10.8. We will have n groups (same as the number of wavelet coefficients for each image), and within each group, there will be N (same as the total number of T_2 weightings) coefficients.

We have argued why the wavelet coefficients should have similar values at similar positions. It is known that the wavelet transform leads to a sparse representation of the individual MR images. If each α_i is approximately s-sparse, then following the argument that different α_i's will have

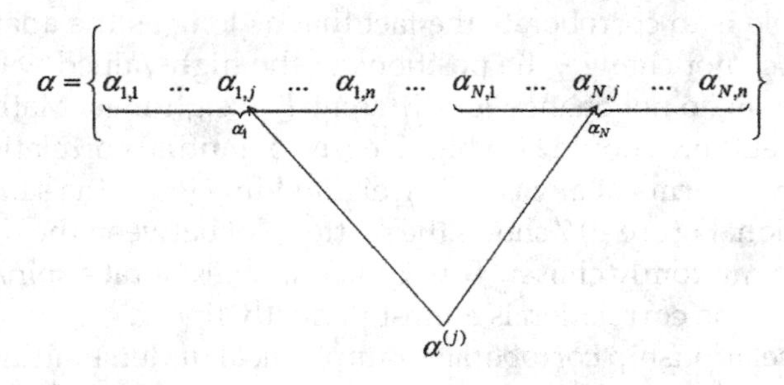

FIGURE 10.8
Grouping of wavelet coefficients according to their position.

high-valued coefficients at similar positions, we can say that the vector α is going to be approximately s-group sparse; that is, there are approximately only s groups that have high-valued wavelet coefficients, and rest of the groups have zero coefficients or coefficients close to zero.

Therefore, one needs to incorporate group sparsity into the optimization problem. We have already studied in previous chapters that $l_{2,1}$ minimization can be used for solving the group sparsity promoting optimization problem,

$$\min_{\tilde{\alpha}} \|\tilde{\alpha}\|_{2,1} \text{ subject to } \|\tilde{y} - \Phi\tilde{\alpha}\|_2 \le \sigma \tag{10.25}$$

where $\|\tilde{\alpha}\|_{2,1} = \sum_{r=1}^{N^2} \left(\sum_{k=1}^{T} \alpha_{k,r}^2 \right)^{1/2}$

The l_2-norm $\left(\sum_{k=1}^{T} \alpha_{k,r}^2 \right)^{1/2}$ over the group of correlated coefficients enforces the selection of the entire group of coefficients, whereas the summation over the l_2-norm enforces group sparsity, that is, the selection of only a few groups.

Group-Sparse Analysis Prior Formulation

The formulation in (10.25) is the synthesis prior formulation. Alternately, one can formulate it as an analysis prior formulation, just as we have seen in the sparse recovery for single-echo MR images. Instead of (10.24), one can express the combined data acquisition model for all the T_2-weighted images as follows,

$$\begin{bmatrix} y_1 \\ \dots \\ y_T \end{bmatrix} = \begin{bmatrix} R_1F & 0 & 0 \\ 0 & \dots & 0 \\ 0 & 0 & R_TF \end{bmatrix} \begin{bmatrix} x_1 \\ \dots \\ x_T \end{bmatrix} + \begin{bmatrix} \eta_1 \\ \dots \\ \eta_T \end{bmatrix} \text{ or, } \tilde{y} = \Psi\tilde{x} + \tilde{\eta} \tag{10.26}$$

where $\tilde{y} = \begin{bmatrix} y_1 \\ \dots \\ y_T \end{bmatrix}, \Psi = \begin{bmatrix} R_1F & 0 & 0 \\ 0 & \dots & 0 \\ 0 & 0 & R_TF \end{bmatrix}, \tilde{x} = \begin{bmatrix} x_1 \\ \dots \\ x_T \end{bmatrix}$ and $\tilde{\eta} = \begin{bmatrix} \eta_1 \\ \dots \\ \eta_T \end{bmatrix}$

Instead of employing a synthesis prior optimization, as in (10.25), one can propose the following group sparsity promoting analysis prior optimization,

$$\min_{\tilde{x}} \|H\tilde{x}\|_{2,1} \text{ subject to } \|\tilde{y} - \Psi\tilde{x}\|_2 \le \sigma \tag{10.27}$$

where $H = \begin{bmatrix} W & 0 & 0 \\ 0 & \dots & 0 \\ 0 & 0 & W \end{bmatrix}$

Just as one obtains better results with analysis prior optimization for single-echo MRI, one gets similar improvements with group-sparse analysis prior as well. This has been verified experimentally.

Experimental Evaluation

The experimental evaluation was carried out on ex-vivo and in-vivo T_2-weighted images of a rat's spinal cord. The data were collected with a 7T MRI scanner. The original ground-truth data consisted of a series of 16 fully sampled echoes acquired with a CPMG sequence with increasing echo time (first echo was acquired with 13.476 ms echo time, and consecutive echoes were acquired with the echo spacing of 13.476 ms). Sixteen T_2-weighted images were collected. In Figure 10.9, echoes 1, 5, 9 and 13 are shown.

Variable density sampling is used; here, the center of the K-space is densely sampled (as most of the signal is concentrated there) and the periphery is sparsely sampled. The sampling ratio is 25%. A third of the total sampling lines is used to densely sample the center of the K-space. The first row of Figure 10.9 shows the ground-truth images. In the subsequent rows, the

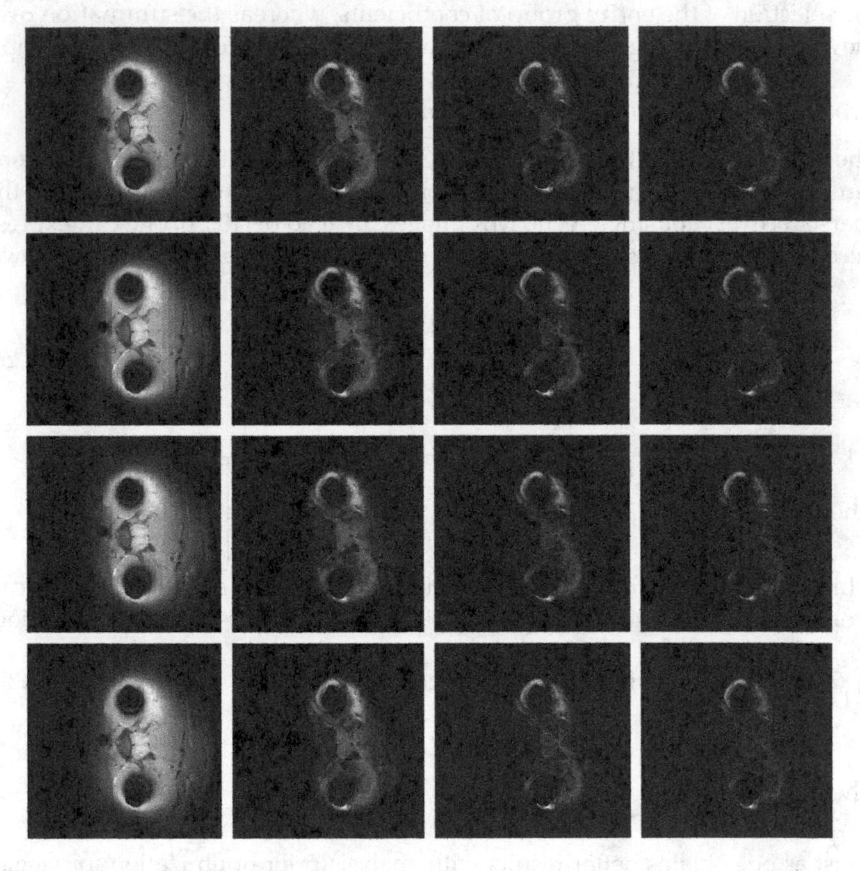

FIGURE 10.9
Top row: ground-truth images. Next three rows: piecemeal reconstruction via CS, group-sparse synthesis prior, and group-sparse analysis prior reconstruction.

reconstruction images are shown. We compare the improvement of group-sparse reconstruction with piecemeal reconstruction of each echo separately.

The improvement in reconstruction quality is clearly discernible from the images. Piecemeal reconstruction of each echo separately leads to blocky artifacts. These blocky reconstruction artifacts are somewhat mitigated in synthesis prior group-sparse reconstruction. They are further mitigated in analysis prior groupsparse reconstruction. Reconstruction via analysis prior group-sparse optimization yields images that closely resemble the ground-truth images generated by full sampling of the K-space.

Multi-coil Parallel Magnetic Resonance Imaging

In the previous section, we learnt how signal processing-based techniques can be employed to accelerate MRI scans. These techniques were developed after the advent of CS. Since these techniques only require changes in the sampling and reconstruction modules of the software, such methods are called software-based acceleration techniques. But prior to the development of such signal processing-based techniques, physics-based acceleration techniques were popular. Physics-based acceleration techniques change the hardware of the scanner in order to facilitate faster scans. Multi-coil parallel MRI is a classic example of physics-based acceleration.

In single-channel MRI, there is a single receiver coil with uniform sensitivity over the full FoV. In multi-channel MRI, there are several receiver coils located at separate locations of the scanner; consequently, they do not have a uniform FoV; for example, see Figure 10.10. Each of the coils can only distinctly "see" a small portion of the full FoV.

In multi-coil MRI, each of the receiver coils partially samples the K-space; since the K-space is only partially sampled, the scan time is reduced. The image is reconstructed from all the K-space samples collected from all coils. The theoretical possibility of combining multi-coil Fourier frequency samples for the purpose of reconstructing a single MR image follows from Papoulis's generalized sampling theorem. The ratio of the total number of possible K-space samples (size of the image) to the number of partial samples collected per coil is called the acceleration factor. In theory, the maximum acceleration factor is the same as the number of coils, but in practice, it is always less than that; for example, if there are 8 receiver coils, the maximum possible acceleration factor will be 8, that is, each coil will sample 12.5% of the total K-space; but in practice, each coil may sample 25% of the K-space, and hence, the acceleration factor will be 4 instead of 8.

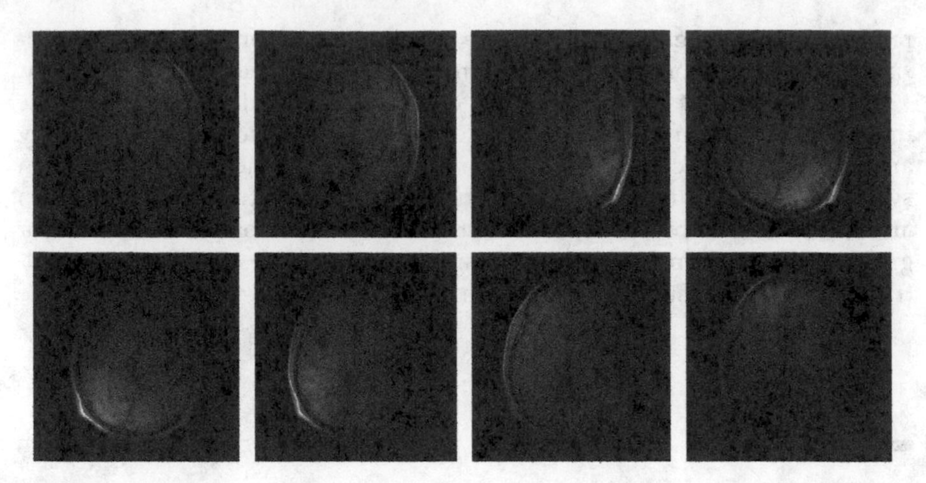

FIGURE 10.10
Images from different coils.

For accelerating the scans, the K-space is partially sampled. There are two possibilities: one can interpolate the unsampled K-space locations, approximate the full K-space, and then reconstruct the image via inverse FFT, or one can directly recover the image from the partially sampled K-space scans. Thus, there are two broad approaches to reconstruct MR images from multi-coil samples:

Frequency domain methods: The unsampled K-space indices are interpolated, from which the image is reconstructed via inverse Fourier transform. There are two broad classes of frequency domain methods. SMASH (Simultaneous Acquisition of Spatial Harmonics (SMASH) and its derivatives use the sensitivity-encoded K-space samples from all coils to estimate the K-space of the original (without sensitivity-encoding) image. Once this is estimated, the image is reconstructed via simple inverse Fourier transform. GeneRalized Autocalibrating Partially Parallel Acquisitions (GRAPPA)-based methods and its variants interpolate the K-space for each of the coils. The sensitivity-encoded coil images are obtained via inverse Fourier transform for each coil. The different coil images are then combined using a sum-of-squares (sos) approach to get the final image. CS techniques are not directly amenable to frequency domain methods, and hence, we will not discuss them any further.

Image domain methods: The image is directly reconstructed (via an unfolding or inversion operation) by combining all the K-space samples from the multiple coils. SENSitivity Encoding (SENSE) and its extensions recover a single image; a new technique called Calibration-less MRI departs from SENSE and recovers the sensitivity-encoded coil images from which the final image needs to be obtained via sos combination. Image domain methods are physically more optimal than frequency domain methods; CS techniques can also be easily integrated. Therefore, we will discuss them in greater detail.

Image Domain Methods

Image domain methods model the data acquisition in a mathematically optimal fashion. There are two image domain methods. SENSE and its associated techniques have been known for long; SENSE directly reconstructs the underlying image. Some recent techniques based on distributed CS reconstruct channel images individually; the final image needs to be computed by sos combination.

SENSitivity Encoding or SENSE

In multi-coil MRI, the coils are situated in different parts of the scanner. Therefore, they do not have a uniform FoV. Thus, they have different sensitivity profiles depending on their FoV. SENSE is an optimal reconstruction technique when the sensitivity profiles are known. Unfortunately, in practice, these are not known and have to be calibrated (from prior scans) or estimated (from the data).

The sensitivity profiles/maps are calibrated based on the assumption that they are smooth; that is, they have only low-frequency components. A low-resolution calibration scan is done before the actual MRI in order to compute the sensitivity maps. Since the maps are smooth, the low-resolution map serves the purpose. However, calibration scans are time-consuming. Additionally, such a protocol assumes that the sensitivity maps remain the same during the calibration scan and during the actual scan—this may not hold true. The more practical way to estimate the sensitivity map is from the data itself. The basic assumption still remains the same; that is, the sensitivity maps are smooth. The sensitivity map is estimated as follows (for each receiver coil):

- For each coil, the center of the K-space of the image is fully sampled, while the rest of the K-space is partially sampled.
- A Kaiser-Bessel window is used to extract the low-frequency components, from which the images are computed by using the 2D inverse Fourier transform.
- The low-frequency images for each coil are normalized by dividing them by the sum-of-squares image. These are used as the sensitivity maps.

So far, we have discussed about estimating sensitivity maps. This is of paramount importance for SENSE reconstruction, since the accuracy of SENSE is largely dependent on the accuracy of the maps. There are other complicated ways to estimate sensitivity maps, and we will not go into them. The two simple methods discussed here are the predominant ones used in practice.

SENSE follows the MRI physical (K-space) data acquisition model. For each coil i, the K-space data acquisition model can be expressed as,

$$y_i = FS_i x + \eta_i, \; i = 1...C \tag{10.28}$$

where y_i is the acquired K-space data of the ith coil, x is the image to be reconstructed, S_i is the diagonalized sensitivity profile (a sensitivity profile is actually a matrix of the same size as the image; in this case, it is converted to a vector by row concatenation, and the diagonal matrix is formed from this vector) of the ith coil, F is the Fourier mapping from the spatial domain to the K-space, η_i is the noise, and C is the total number of coils.

In a compact matrix vector form, (10.28) can be expressed as,

$$y = Ex + \eta \tag{10.29}$$

where $y = \begin{bmatrix} y_1 \\ \cdots \\ y_C \end{bmatrix}$, $E = \begin{bmatrix} FS_1 \\ \cdots \\ FS_C \end{bmatrix}$ and $\eta = \begin{bmatrix} \eta_1 \\ \cdots \\ \eta_C \end{bmatrix}$.

Here, E is called the encoding operator. SENSE reconstruction solves the inverse problem (10.28) and thereby obtains the image x. This is an optimal method when accurate values of the sensitivity profiles S_i's are known.

Traditionally, the image is estimated via least squares minimization, that is,

$$\hat{x} = \min_x \|y - Ex\|_2^2 \tag{10.30}$$

Regularized SENSE

Least squares minimization yields the best results. We have already discussed the problems associated with simple least squares regularization. To prevent these issues, generally, the SENSE reconstruction is solved with a regularization term,

$$\hat{x} = \min_x \|y - Ex\|_2 + \lambda Z(x) \tag{10.31}$$

where λ is a regularization parameter (a scalar) and $Z(x)$ is the regularization term. The regularization term can be Tikhonov regularization, TV regularization, or wavelet regularization. TV and wavelet regularizations produce MR images with sharp edges and are the precursors of CS-based techniques. In general, the regularization terms capture some prior information regarding the nature of the reconstructed image.

Tikhonov regularization is a well-known technique in least squares regression. It has a nice closed-form solution; however, it does not provide any insight into the nature of the solution and is never used in practice. The often-used regularization technique is TV. It assumes that the image is piecewise linear. Therefore, the gradient of the image will be sparse. We have discussed about TV previously, but we refresh our memory here. The TV is defined as,

$$TV(x) = \Sigma\sqrt{(D_h x)^2 + (D_v x)^2}$$

where D_h and D_v are horizontal and vertical differentiation operators, respectively.

The regular TV regularization assigns equal weights to all the derivatives. In recent years, non-local techniques have become popular in image processing. Non-local techniques adaptively vary the weights of the operator, so that the salient features of the image are not destroyed. The weights are patch-based instead of being point-based in order to be robust.

The NLTV regularization term is defined as,

$$NLTV(x) = \sum \sqrt{w(m,n)}(x_m - x_n) \tag{10.32}$$

Here, m and n are two points in the image, and the weight is defined as,

$$w(m,n) = \frac{\tilde{w}(m,n)}{\sum \tilde{w}(m,n)}$$

$$\tilde{w}(m,n) = \begin{cases} \exp\left(-\frac{\|p(m)-p(n)\|_2^2}{2\sigma^2}\right) & \text{if } \|m-n\| \leq \delta/2 \\ 0 & \text{otherwise} \end{cases} \tag{10.33}$$

Here, $p(m)$ denotes the patch around m, δ controls the non-locality (if the points are too far, they are not considered, i.e., the weight is 0), and σ controls the spread of the patch and needs to be set, depending on noise level.

Experimental Evaluation

In-vivo brain data from a healthy volunteer was acquired on a 3T commercial scanner with an 8-channel head coil. The reconstruction results for an acceleration factor of 4 are shown in Figure 10.11. It is easy to notice that NLTV has better reconstruction results than SENSE (CG in Figure 10.11), Tikhonov-regularized SENSE, and TV-regularized SENSE. The aliasing artifacts are hardly visible in NLTV-regularized SENSE reconstruction.

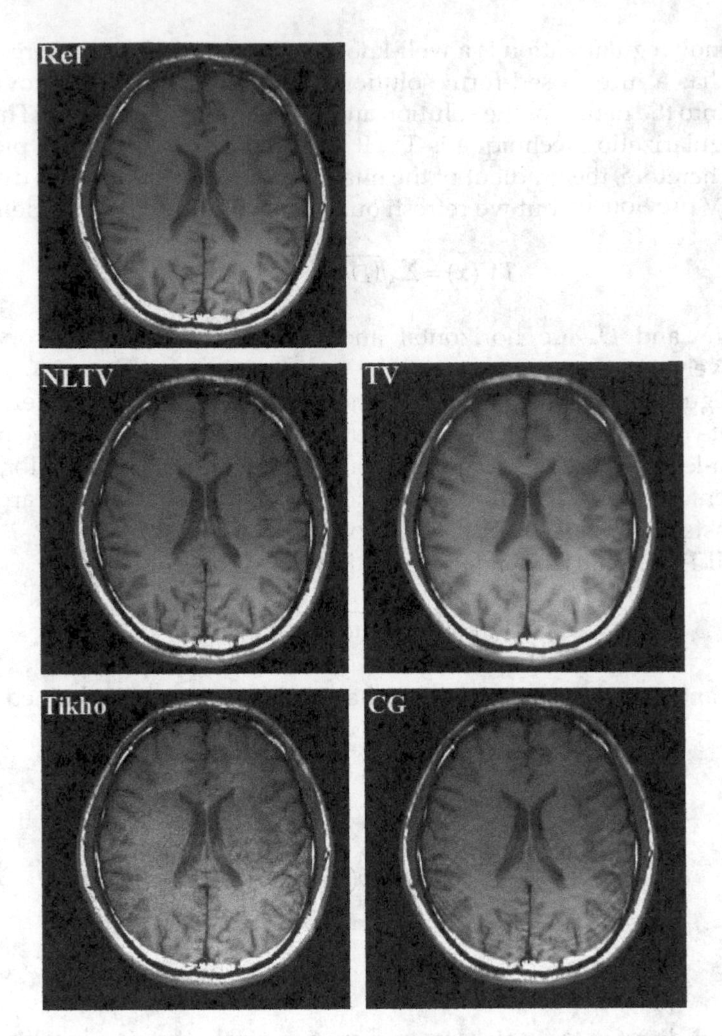

FIGURE 10.11
The reconstructions using four different methods with reduction factor 4 from an 8-channel scanned human brain data.

Compressed Sensing SENSE

Previously, we learnt how CS can be used for reconstructing single-channel MR images. The same formulation can be directly borrowed for SENSE reconstruction, with the only difference that, instead of the simple Fourier operator, we have a SENSE-encoding operator; everything else remains the same.

As in single-channel reconstruction, the problem here is also formulated in the wavelet domain. Since wavelet coefficients of MR image are sparse, these coefficients can be retrieved by solving the following problem,

$$\hat{\alpha} = \min_{\alpha} \|\alpha\|_1 \text{ subject to } \|y - EW^T\alpha\|_2^2 \leq \varepsilon \tag{10.34}$$

where ε is dependent on the noise and α is the wavelet coefficient vector.

All previous studies that leveraged CS in addressing the SENSE problem have solved an over-determined system of equations. Thus, in the true sense of the term, problem (10.34) is not a CS problem; this is because it is an over-determined problem (E is over-determined), and the l_1-norm over the wavelet coefficients acts as a regularization term. Thus, in effect, the CS SENSE reconstruction does not yield significantly different reconstruction from regularized SENSE.

Iterative SENSE

All the aforementioned studies assume that the sensitivity maps are known. The accuracy of the reconstruction technique depends on the provided sensitivity maps. However, the estimates of the sensitivity maps are not always accurate. In order to get a robust reconstruction, a joint estimation of the image and sensitivity map has been proposed. The SENSE equation is expressed as,

$$y = FS(a)x + \eta \tag{10.35}$$

where $S(a) = \begin{bmatrix} S_1(a) \\ \cdots \\ S_C(a) \end{bmatrix}$

The sensitivity map for each coil is parameterized as a polynomial of degree N, that is, the sensitivity for the ith coil at position (m, n) is given by $s_i(m,n) = \sum_{p=0}^{N} \sum_{q=0}^{N} a_{i,p,q} x^p y^q$, where (m, n) denotes the position and N and a denote the degree of the polynomial and its coefficients, respectively.

In order to jointly estimate the coil sensitivities (given the polynomial formulation) and the image, the following problem has to be solved,

$$\arg\min_{a,x} \|y - RFS(a)x\|_2 \tag{10.36}$$

However, it has been argued that solving for all the variables jointly is an intractable non-convex problem; hence, the image and the polynomial coefficients were solved iteratively, as follows:

$$\arg\min_{x} \|y - RFS(a)x\|_2 \tag{10.37}$$

The image is reconstructed assuming that the coefficient a is fixed. Then,

$$\arg\min_{a} \|y - RFS(a)x\|_2 \tag{10.38}$$

The sensitivity map is estimated assuming that the image x is fixed.

This method for solving the parallel MRI reconstruction problem via iteratively solving (10.37) and (10.38) is called Joint Sensivity Encoding (JSENSE). Even though (10.37) and (10.38) are convex problems, the original problem (10.36) is non-convex, and thus, there is no guarantee that solving it by iterating between the steps will converge to the desired solution. However, with reasonably good initial estimates of the sensitivity maps, JSENSE provides considerable improvement over SENSE reconstruction.

The JSENSE can be regularized using the CS framework. It is assumed that the image is sparse in a finite difference domain and the coil sensitivity is sparse in Chebyshev polynomial basis or in Fourier domain. This leads to,

$$\hat{x} = \arg\min_{x} \|y - RFSx\|_2 + \lambda_x TV(x) \tag{10.39}$$

$$\hat{S} = \arg\min_{S} \|y - RFX_S\|_2 + \lambda_s \Psi \|s\|_1 \tag{10.40}$$

In (10.39), the image is solved assuming that sensitivity map is known. In (10.40), the sensitivity map is estimated assuming that the image is known. Here, Ψ is the sparsifying basis (Chebyshev polynomial basis or Fourier basis) for the sensitivity maps.

In a previous chapter, we learnt about the incoherency requirement in CS; in order for CS to be successful, the measurement basis and the sparsifying basis need to be incoherent. However, in the CS-regularized JSENSE formulation, Chebyshev basis is used for sparsifying the sensitivity maps, and the measurement basis is Fourier. Chebyshev sparsifying basis and Fourier measurement basis are highly coherent. Thus, Chebyshev basis is not a theoretically good choice for CS-based reconstruction in the current context.

The iterative SENSE (iSENSE) proposed a solution to this problem. They proposed to model the sensitivity maps as rank-deficient matrices. This is a new assumption. In order to corroborate it empirically, we show plot of singular-value decay of the three sensitivity maps (one, three, and five) for an 8-channel MRI scanner in Figure 10.12. It shows that the singular values of the sensitivity maps decay fast, and thus, the sensitivity maps are approximately rank-deficient.

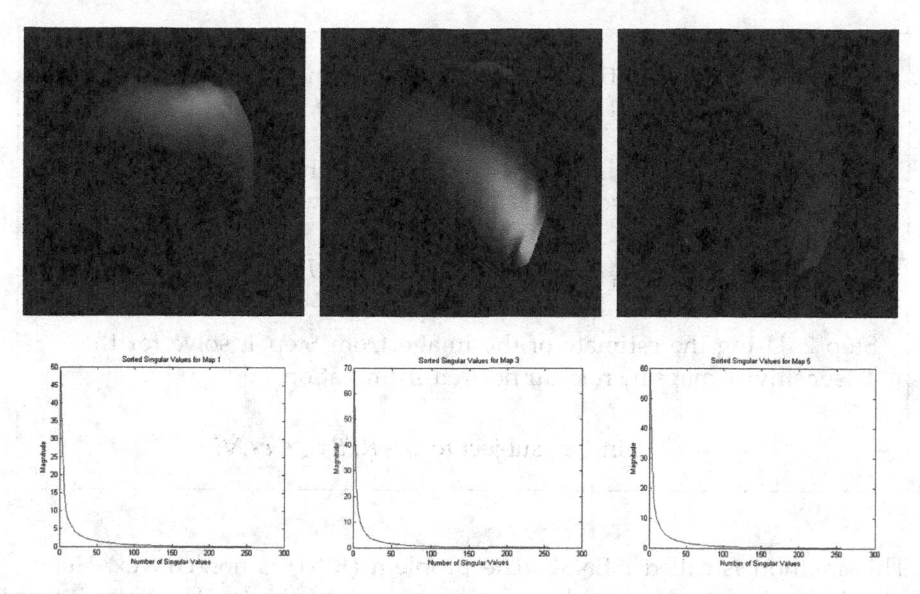

FIGURE 10.12
Top row: sensitivity maps. Bottom row: Decay of singular values.

The rank deficiency of the sensitivity maps can be incorporated into the CS-based MR image reconstruction problem. Furthermore, in order to exploit the sparse nature of the transform coefficients of the MRI image, the reconstruction problem was formulated as,

$$\begin{aligned} &\min_{x,S} \left\| y - RFS_x \right\|_2 \text{ subject to} \\ &\left\| Wx \right\|_1 + \gamma TV(x) \le \tau_1, \\ &\left\| S_i \right\|_* \le \tau_2, \forall i \end{aligned} \tag{10.41}$$

where τ_1 and τ_2 are the constraints on the transform domain sparsity of the image and on the rank deficiencies of the sensitivity maps, respectively, and γ is the term balancing the relative importance of the transform domain sparsity and TV.

The problem in (10.39 and 10.40) is that since sparsity is used for estimating the sensitivity maps, the estimation requires incoherence between the measurement (Fourier) and sparsifying (Chebyshev) bases. Here, rank deficiency is exploited for estimating the sensitivity maps; therefore, one does not encounter the problem arising out of the incoherency criterion.

Similar to JSENSE, (10.41) is a non-convex problem. Solving for the MR image and the sensitivity maps jointly is intractable. Therefore, following the approach in JSENSE, an iterative solution was proposed.

Initialize: An initial estimate of the sensitivity maps.
Iteratively solve Steps 1 and 2 until convergence.

Step 1: Using the initial or the estimates of the sensitivity maps obtained in Step 2, solve for the image by solving a CS problem,

$$\min_x \|Wx\|_1 + \gamma TV(x) \text{ subject to } \|y - RFSx\|_2 \leq \varepsilon_1$$

Step 2: Using the estimate of the image from Step 1, solve for the sensitivity maps by nuclear norm minimization,

$$\min_S \|Sx\|_* \text{ subject to } \|y - RFSx\|_2 \leq \varepsilon_2, \forall i$$

This method is called iSENSE. The problem (10.41) is non-convex. Hence, its solution is not guaranteed to reach a unique minimum. However, it was found that, practically, the said algorithm converges toward three iterations.

Experimental Evaluation

In Figure 10.13, one can see how the iSENSE reconstruction improves in subsequent iterations. The top portion of each image has been magnified for better clarity. It is easy to see from the magnified portions after the first step that there are some fine anatomical details that are missing. After the second iteration, some of the details are captured. The third iteration shows slight improvement over the second iteration. By the third iteration, the anatomical detail captured is as good as the ground-truth image. It should be noted that the first step of the iteration is the same as CS SENSE. Thus, the iSENSE greatly improves over CS SENSE in three iterations.

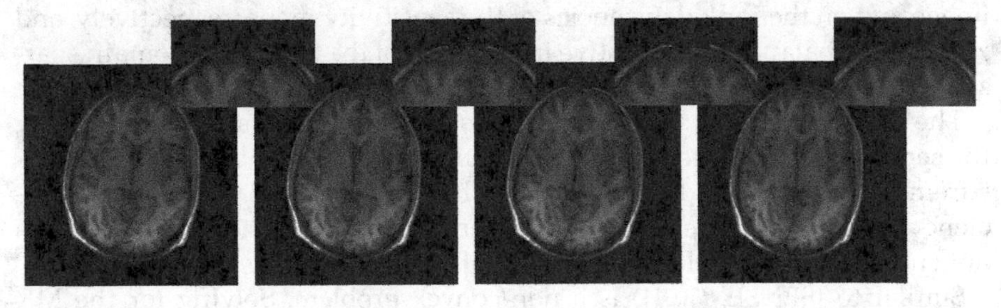

FIGURE 10.13
iSENSE reconstructed images in each step.

Calibration-Less Magnetic Resonance Imaging Reconstruction

All multi-coil parallel MR image reconstruction algorithms require some form of calibration. The image domain methods require the sensitivity profiles to be estimated from the calibration data, while the frequency domain methods require computing the linear interpolation weights from the calibration data. In both cases, the assumption is that the estimates based on the calibration data hold for the unknown image (to be reconstructed) as well. For image domain methods, there are different ways to compute the sensitivity profiles, and the reconstruction results are dependent on the technique used to compute them. For frequency domain methods, the interpolation weights are dependent both on the calibration data and on the calibration kernel. In short, all known methods for multi-coil parallel MRI are dependent on the calibration stage.

The iSENSE and the JSENSE techniques are more robust than SENSE. Such iterative schemes improve robustness to initial calibration. However, these do not offer the perfect solution; the final images are still sensitive to the estimation of the sensitivity maps or interpolation weights. Ideally, we need a technique that does not require calibration at all.

In order to alleviate the sensitivity issues associated with traditional image domain and frequency domain methods, a new technique was proposed; it does not require calibration or sensitivity estimation in any form and is dubbed as Calibration-Less Multi-coil (CaLM) MRI.

One can rewrite the physical data acquisition model for multi-coil MRI as,

$$y_i = FS_i x + \eta_i, i = 1...C$$

This can be expressed in a slightly different fashion in terms of the sensitivity-encoded images,

$$y_i = Fx_i + \eta_i, i = 1...C \tag{10.42}$$

where $x_i = S_i x$ are the sensitivity-encoded images for each coil.

By incorporating the wavelet transform into (10.42), one gets,

$$y_i = FW^T \alpha_i + \eta_i, i = 1...C \tag{10.43}$$

Following the works in CS-based MR image reconstruction, one can reconstruct the individual coil images separately by exploiting their sparsity in some transform domain; that is, each of the images can be reconstructed by solving,

$$\min_{\alpha_i} \|\alpha_i\|_1 \text{ subject to } \|y_i - FW^T\alpha_i\|_2^2 \le \varepsilon_i \tag{10.44}$$

This is the synthesis prior formulation; it solves for the sparse transform coefficients. In situations where the sparsifying transform is not orthogonal[4] or a tight frame,[5] the inverse problem (10.42) can be alternately solved via the following analysis prior optimization,

$$\min_{x_i} \|Wx_i\|_1 \text{ subject to } \|y_i - Fx_i\|_2^2 \le \varepsilon_i \quad (10.45)$$

However, such piecemeal reconstruction of coil images (be it analysis or synthesis) does not yield optimal results. The data acquisition model can be expressed in the multiple measurement vector (MMV) form as follows,

$$Y = FX + N \quad (10.46)$$

where $Y = [y_1|\ldots|y_C]$, $X = [x_1|\ldots|x_C]$ and $N = [\eta_1|\ldots|\eta_C]$.

Here, the problem is to recover X.

The multi-coil images (x_is) are formed by sensitivity encoding of the original image (to be reconstructed). All previous studies in parallel MRI assume that the sensitivity maps are smooth and have a compact support in the Fourier domain. CaLM MRI postulates that the sensitivity maps are smooth and hence do not alter the positions of the edges of the images, although they might change the absolute values.

This can be clarified with a toy example. Figure 10.14a shows a prominent edge (say after sensitivity encoding by first coil), and Figure 10.14b shows a less prominent edge (say after sensitivity encoding by second coil).

If finite difference is used as the sparsifying transform, the discontinuities along the edges are captured; that is, there are high values along the edges but zeroes elsewhere. The positions of the discontinuities are maintained, but the absolute values change, as can be seen from Figure 10.15.

Based on this toy example, we consider the MMV formulation (10.46). All the columns of X are images corresponding to different coils. Since the sensitivity maps of all the coils are smooth, the positions of the edges remain intact. For better clarity, we look at the images in a transform domain,

1	1	0	0
1	1	0	0
1	1	0	0

(a)

0.5	0.5	0	0
0.5	0.5	0	0
0.5	0.5	0	0

(b)

FIGURE 10.14
(a) Sharp edge, and (b) Less prominent edge.

[4] Orthogonal: $\Psi^T\Psi = I = \Psi\Psi^T$
[5] Tight-frame: $\Psi^T\Psi = I \neq \Psi\Psi^T$

0	1	0	0		0	0.5	0	0
0	1	0	0		0	0.5	0	0
0	1	0	0		0	0.5	0	0
	(a)					(b)		

FIGURE 10.15
(a) Finite differencing of sharp edge, and (b) finite differencing of less prominent edge.

$$WX = Z = \begin{bmatrix} \alpha_{1,1} & \cdots & \alpha_{1,C} \\ \cdots & \cdots & \cdots \\ \alpha_{r,1} & \cdots & \alpha_{r,C} \\ \cdots & \cdots & \cdots \\ \alpha_{N,1} & \cdots & \alpha_{N,C} \end{bmatrix} \quad (10.47)$$

where W is the sparsifying (wavelet) transform that encodes the edges of the images and Z is the matrix formed by stacking the transform coefficients as columns.

In (10.47), each row corresponds to one position. Based on the discussion so far, since the positions of the edges in the different images do not change, the positions of the high-valued coefficients in the transform domain do not change either. Therefore, for all the coil images, the high-valued transform coefficients appear at the same position. Thus, the matrix Z is row-sparse; that is, there are a few rows with high-valued coefficients, while most of the rows are zeros.

Based on this discussion, the MMV recovery problem (10.46) can be solved by exploiting this row-sparsity information. The analysis prior formulation is as follows,

$$\min_X \|WX\|_{2,1} \text{ subject to } \|Y - FX\|_F^2 \leq \varepsilon \quad (10.48)$$

In situations where the sparsifying transform is orthogonal or a tight frame, the inverse problem can be solved via the following synthesis prior optimization,

$$\min_Z \|Z\|_{2,1} \text{ subject to } \|Y - FW^T Z\|_F^2 \leq \varepsilon \quad (10.49)$$

where $Z = WX$.

The coil images can be recovered by applying the synthesis equation: $X = \Psi^T Z$.

The final image (I) is obtained from the individual coil images by sum-of-squares combination in a fashion similar to GeneRalized Autocalibrating Partial Parallel Acquisition (GRAPPA) and SPIRiT, that is, $I = \left(\sum_{i=1}^{C} x_i^2\right)^{1/2}$.

Experimental Evaluation

Experimental results from three datasets are shown in Figure 10.16. The brain data is a fully sampled T1-weighted scan of a healthy volunteer. The scan was performed on a GE Sigma-Excite 1.5-T scanner, using an eight-channel receiver coil. The eight-channel data for Shepp-Logan phantom was simulated. The UBC MRI Lab prepared a phantom by doping saline with gadolinium-diethylenetriamine pentaacetic acid (Gd-DTPA). A four-coil scanner was used for acquiring the data by using a Fast Low Angle Shot (FLASH) sequence (Figure 10.16).

From the reconstructed images, it is obvious that CaLM MRI yields better reconstruction than the commercial and state-of-the-art methods. This is best visible from the Shepp-Logan phantom. The reconstruction artifacts are more pronounced in GRAPPA and CS SENSE.

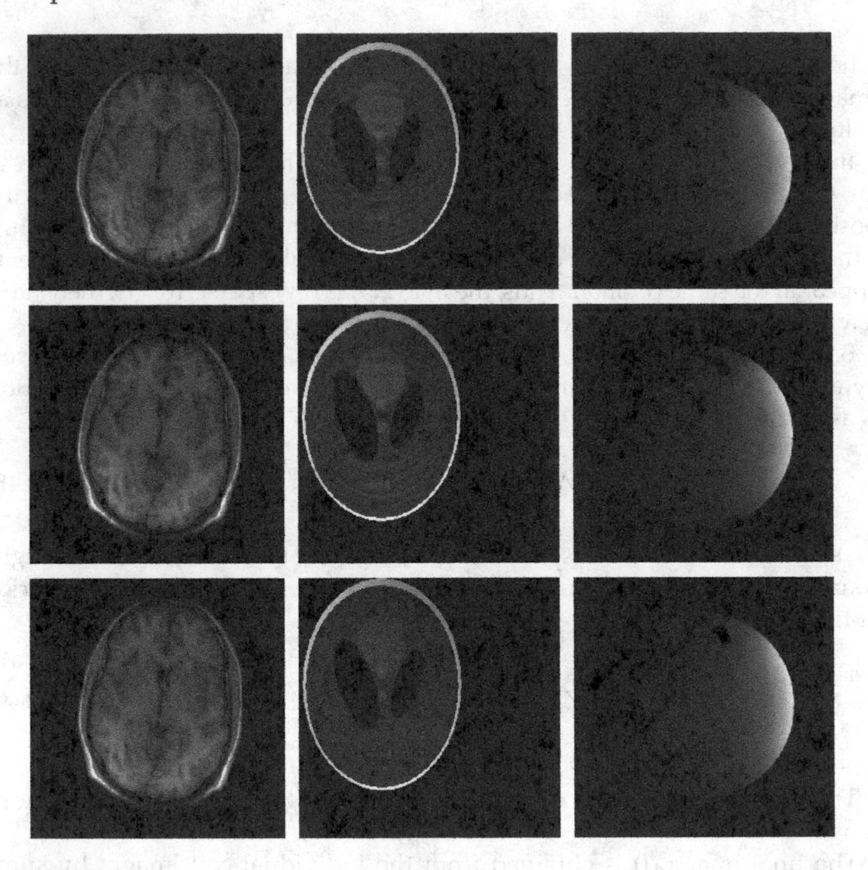

FIGURE 10.16
Reconstructed images from various methods. From top to bottom: GRAPPA, CS SENSE, CaLM synthesis prior, and CaLM analysis prior.

Dynamic Magnetic Resonance Imaging Reconstruction

So far, we have been studying static MRI reconstruction. We were interested in the structure and anatomy of the cross-section. In dynamic MRI, our interest is to study changing behavior—we study motion such as blood flow and heartbeat, or changes in concentration, such as blood oxygen level dependent (BOLD) signals. If we use a simile from digital photography, static MRI is a photo and dynamic MRI is a video.

The problems that fraught static MRI are pertinent here as well. We would like to accelerate dynamic MRI scans to the maximum possible extent. This is not only to reduce patient discomfort but also for technical reasons. Ideally, we would like to acquire a dynamic MRI video with high spatial and temporal resolutions. However, we know that since the number of K-space samples that can be acquired in a unit of time is limited, there is always a trade-off between spatial and temporal resolutions. Therefore, by traditional means, we can only get scans of high spatial resolution with low temporal resolution or vice versa. Such scans do not allow detection of fast events (owing to limitations in temporal resolution) or events concentrated in small areas (owing to limitations in spatial resolution).

In conventional dynamic MRI scans, the K-space for each frame is fully sampled. The frames are reconstructed by inverse Fourier transform. We have learnt in earlier chapters how CS has helped to accelerate static MRI scans. Similar techniques will be applicable here.

The dynamic MRI data acquisition model can be expressed succinctly. Let x_t denote the MR image frame at the tth instant. We assume that the images are of size N × N and T is the total number of frames collected. Let y_t be the K-space data for the tth frame. The problem is to recover all x_ts ($t = 1 \ldots T$) from the collected K-space data y_ts. The MR imaging equation for each frame is as follows,

$$y_t = RFx_t + \eta \tag{10.50}$$

where R is the sub-sampling mask, F is the forward Fourier operator, and η is white Gaussian noise.

In dynamic MRI, each of the frames is spatially correlated—this follows from our knowledge in static MRI reconstructed. Thus, these frames are sparse in transform domains such as wavelets and also have small TVs. One should also notice that the dynamic MRI sequence is temporally correlated; this is because temporal changes are slow, and one frame looks much like the previous ones. Thus, in dynamic MRI, the inter-frame temporal correlation should also be exploited along with intra-frame spatial correlations, in order to achieve the best possible reconstruction.

k-t FOCUSS (k-t BLAST)

One of the earliest studies in dynamic MRI reconstruction drew inspiration from video processing. It models dynamic imaging; thus,

$$x_t = x_{ref} + x_{diff} \tag{10.51}$$

The tth frame (x_t) is expressed as a deviation (x_{diff}) from some reference frame (x_{ref}).

Since the frames are correlated, the difference frame will be sparse. Drawing inspiration from video coding, they further argued that the difference frame will be even sparser in some transform domain. Instead of reconstructing the tth frame, they proposed to reconstruct the difference instead,

$$\hat{x}_{diff} = \min_{x_{diff}} \left\| y_t - RFx_{ref} - RFx_{diff} \right\|_2^2 + \lambda_1 \left\| Wx_{diff} \right\| + \lambda_2 TV(x_{diff}) \tag{10.52}$$

Here, W denotes the wavelet transform and TV denotes total variation.

Transform Domain Method

There is a straighter approach to CS-based dynamic MRI reconstruction. One can assume that the wavelet transform sparsifies along the spatial dimension and Fourier transform sparsifies along the temporal dimension; since both these studies were experimenting on cardiac data (which is semi-periodic), this was a reasonable assumption. The offline dynamic MRI acquisition model can be expressed in the following manner,

$$Y = RFX + \eta \tag{10.53}$$

where $Y = \left[y_1 | \ldots | y_T \right]$, $X = \left[x_1 | \ldots | x_T \right]$ and $\eta = \left[\eta_1 | \ldots | \eta_T \right]$.

The standard CS optimization problem can now be employed,

$$\hat{X} = \min_X \left\| W \otimes F_{1D}(X) \right\|_1 \text{ subject to } \left\| Y - RFX \right\|_F^2 \leq \varepsilon \tag{10.54}$$

where $W \otimes F_{1D}$ is the Kronecker product[6] of the wavelet transform (for sparsifying in space) and F_{1D} is the 1D Fourier transform (for sparsifying along the temporal direction).

[6] Let $A_{m \times n}$ and $B_{p \times q}$ be two matrices and

$$A = \begin{bmatrix} a_{11} & \cdots & a_{1n} \\ \cdots & \cdots & \cdots \\ a_{m1} & \cdots & a_{mn} \end{bmatrix}, \text{ then } A \otimes B = \begin{bmatrix} a_{11}B & \cdots & a_{1n}B \\ \cdots & \cdots & \cdots \\ a_{m1}B & \cdots & a_{mn}B \end{bmatrix}$$

Experimental Evaluation

Some experiments were carried out on cardiac imaging data. A short-axis cine dataset was acquired on a Philips 3.0T Achieva system. The ground-truth data comprised the fully sampled K-space, from which the undersampling patterns were simulated for different acceleration factors. CS reconstruction was compared against k-t Broad-use Linear Acquisition Speed-up Technique (BLAST), which is a well-known dynamic reconstruction technique from the yesteryears. Portion of a heart's phases 1 and 4 are shown in Figure 10.17.

In Figure 10.17, the difference frames are contrast enhanced 10 times for better visual clarity. For k-t BLAST, the difference image shows higher errors along the edges of the image; this implies bad reconstruction. For CS reconstruction, the difference is evenly distributed over the image; the errors do not concentrate along edges and are nearly independent of the underlying structure of the image. This is good news—reconstruction error should be error-like and should not be correlated with the structure of the image. This allows further improvement in quality by denoising the reconstructed image, but if the reconstruction error has structure, the image cannot be improved by simple denoising.

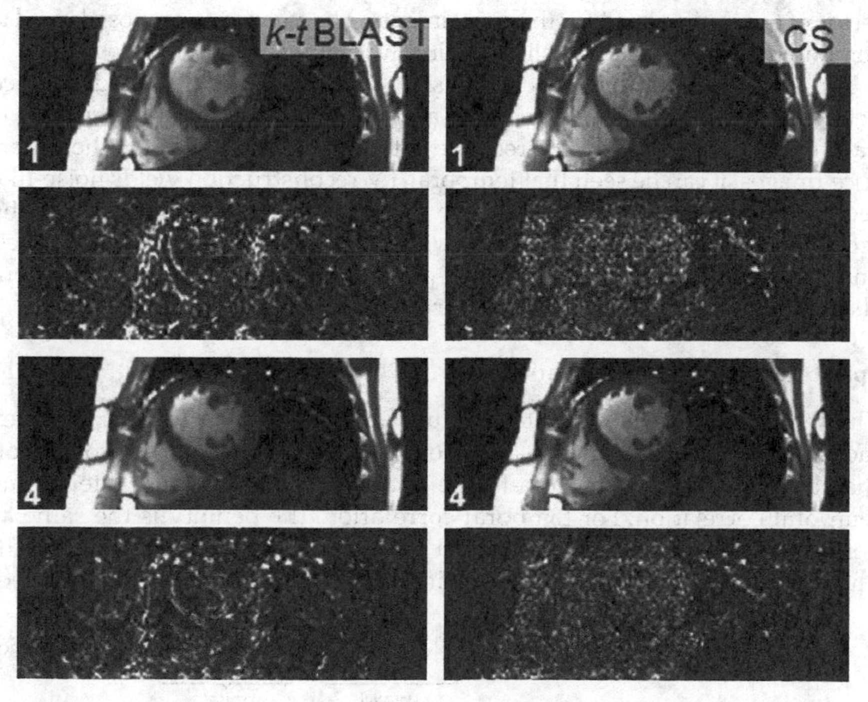

FIGURE 10.17
Reconstructed and difference frames.

Temporal Total Variation

A simpler way to implicitly model the spatial redundancy among the frames is to only account for the difference between frames; since the frames are temporally correlated, the difference will be sparse.

$$\hat{X} = \min_X \|Y - RFX\|_{Fro}^2 + \lambda \left\| \sqrt{\nabla_t(X)^2} \right\|_1 \quad (10.55)$$

Here, $\nabla_t(X)^2 = \sum_t \frac{d}{dt}(x_t)^2$ denotes the TV along the temporal direction. The assumption is that, if the frames are temporally correlated, the difference between subsequent frames will be sparse and hence the temporal TV will be small.

Experimental Evaluation

For the experiments, breast dynamic contrast enhanced magnetic resonance imaging (DCE-MRI) data were acquired from a volunteer having two tumor. Omniscan of dose 0.1 mL/kg was injected at 4 mL/s, followed by 20 mL saline flush injected at 2 mL/s. Temporal resolution per frame was 12–15. The ground-truth data consisted of the fully acquired K-space, from which an acceleration factor of 6 was simulated for the experiments. The technique proposed in [5] is compared with the standard sliding window method.

The 12th frame of the sequence is shown in Figure 10.18. The difference between temporal TV reconstruction and sliding window reconstruction is hard to be discerned from the reconstructed images. However, in the difference images, it can be seen that temporal TV reconstruction yields noise-like reconstruction artifacts, which can be easily removed by denoising. But the reconstruction artifacts from sliding window technique are more structured and hence cannot be removed by post-processing. Thus, reconstruction from sliding window technique is of poorer quality than TV reconstruction.

Spatio-temporal Total Variation

The temporal TV method is too simplistic and will not hold for complex motion. Thus, it is not advisable to use it in practice. One can improve upon that by exploiting both intra-frame spatial correlation and inter-frame temporal correlation. For temporal correlation, the penalty is the same as before—temporal TV. But for spatial correlation, the penalty is the well-known spatial TV. The following optimization problem needs to be solved,

$$\hat{X} = \min_X \|Y - RFX\|_{Fro}^2 + \underbrace{\lambda \left\| \sqrt{\nabla_t(X)^2} \right\|_1}_{temporal} + \underbrace{\lambda \left\| \sqrt{\nabla_s(X)^2} \right\|_1}_{spatial} \quad (10.56)$$

Such a formulation can solve a wider variety of problems than (10.55).

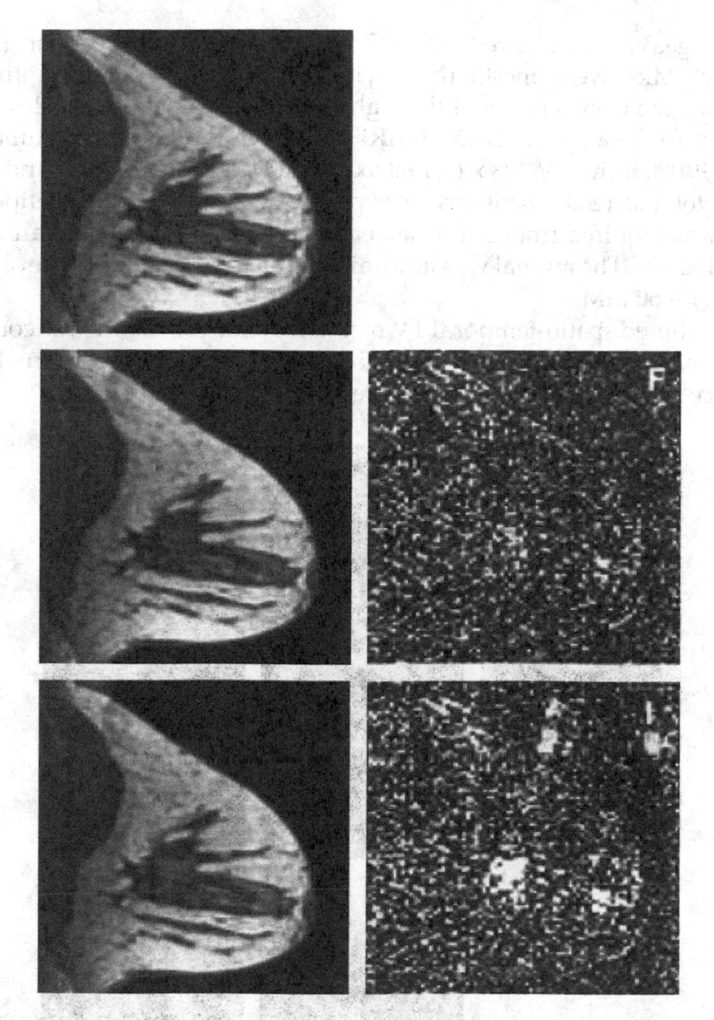

FIGURE 10.18
Reconstructed images. Top to bottom: ground-truth, temporal TV reconstruction, and sliding window reconstruction.

Experimental Evaluation

DCE-MRI experiments were performed on female tumor-bearing non-obese diabetic/severe combined immune-deficient mice. All animal experimental procedures were carried out in compliance with the guidelines of the Canadian Council for Animal Care and were approved by the institutional animal care committee. Tumor xenografts were implanted subcutaneously on the lower-back region.

All images were acquired on a 7T/30-cm bore MRI scanner (Bruker, Germany). Mice were anesthetized with isofluorane, and temperature and respiration rate were monitored throughout the experiment. FLASH was used to acquire fully sampled 2D DCE-MRI data from the implanted tumor with 42.624 × 19.000 mm FOV, 128 × 64 matrix size, TR/TE = 35/2.75 ms, and 40° flip angle. A total of 1200 repetitions were performed at 2.24 s per repetition. The dataset was acquired from a mouse bearing HCT-116 tumor (human colorectal carcinoma). The animal was administered 5 μL/g Gadovist® (Leverkusen, Germany) at 60 mM.

The combined spatio-temporal TV reconstruction technique was compared with the temporal TV method. The reconstruction results (reconstructed and difference images) are shown in Figure 10.19.

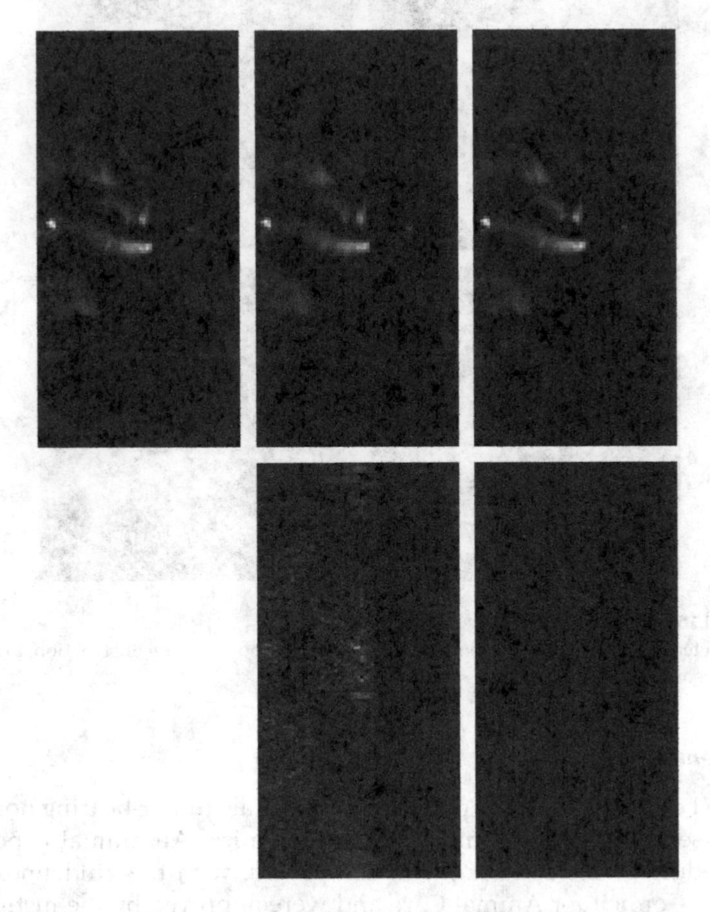

FIGURE 10.19
Reconstructed (top row) and difference images (bottom row). From left to right: ground-truth, temporal TV, and spatio-temporal TV.

The difference images in Figure 10.19 have been contrast-enhanced by 5 times for better visual clarity. From the difference images, it can be easily seen that the spatio-temporal TV yields far superior reconstruction than temporal TV.

Low-Rank Methods in Dynamic Magnetic Resonance Imaging Reconstruction

CS-based techniques are the de facto standards in offline dynamic MRI reconstruction. However, a few studies proposed an alternating approach based on low-rank matrix recovery techniques.

The matrix X is formed by stacking the frames as its columns. The frames are correlated; therefore, the columns of the matrix are linearly dependent, and hence, the matrix X is low-rank. Based on this argument, one can solve the following inverse problem by using the assumption that the solution is low-rank.

$$Y = RFX + \eta$$

We have studied in a previous chapter that it is indeed possible to recover low-rank matrices from their lower-dimensional projections. Thus, it is possible to recover X by solving the following nuclear norm minimization problem,

$$\hat{X} = \min_X \|X\|_* \text{ subject to } \|Y - RFX\|_2^2 \le \varepsilon \tag{10.57}$$

Here, we are showing a nuclear norm minimization-based approach, since the problem is convex and well-known guarantees exist, but one can also solve the same by using matrix factorization techniques.

Experimental Evaluation

Simulations were performed using a cardiac cine dataset, which was spatiotemporally warped to emulate a 5-second free-breathing experiment in the presence of cardiac arrhythmia. The ground truth is shown in Figure 10.20a. An acceleration factor of 5 was simulated by randomly sampling the k-t space (showed in Figure 10.20b). Reconstruction is carried out using the low-rank recovery model (Figure 10.20c) and by simple zero filling (Figure 10.20d). It is easy to see that the reconstruction results from the low-rank recovery model are better.

The use of low-rank recovery techniques for dynamic MRI reconstruction is a proof-of-concept but not a competing reconstruction method. There is no evidence that it will perform as good as CS-based reconstruction.

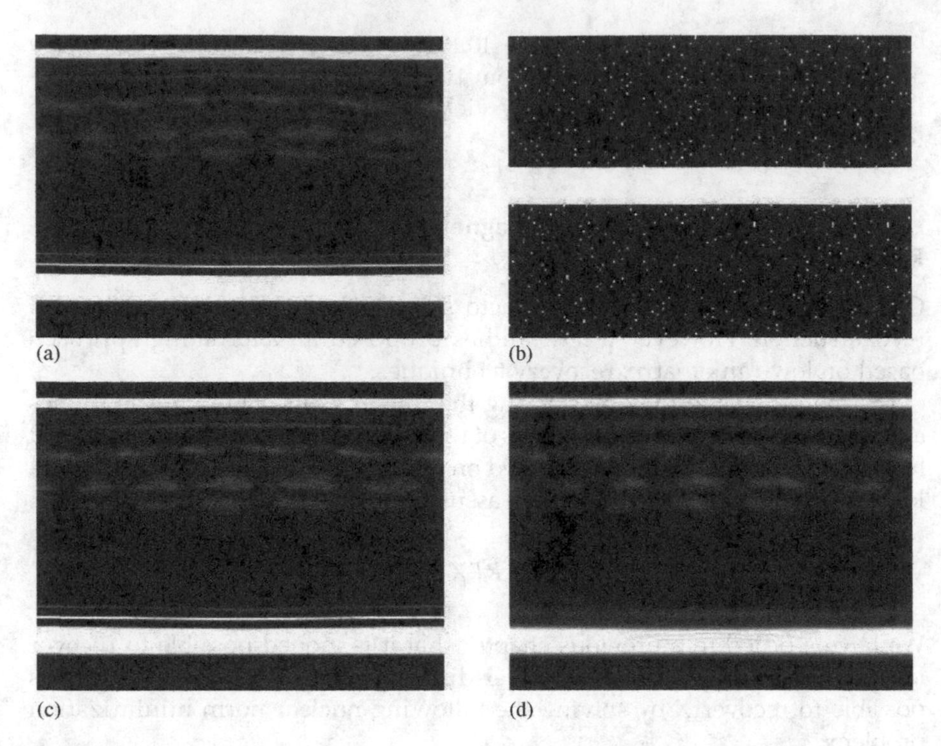

FIGURE 10.20
Reconstructed images by low-rank techniques. (a) Groundtruth (fully sampled). (b) Random Sampling Pattern. (c) Zero-filled Reconstruction. (d) Low rank reconstruction.

Combined Low-Rank and Sparsity-Based Techniques

CS-based techniques maximally exploit the spatio-temporal redundancy in the dynamic MRI sequences in order to reconstruct them. A typical such formulation is (10.54), which is repeated here for the sake of completeness,

$$\hat{X} = \min_X \left\| W \otimes F_{1D}(X) \right\|_1 \text{ subject to } \left\| Y - RFX \right\|_{Fro}^2 \leq \varepsilon \tag{10.58}$$

On the other hand, low-rank recovery methods suggest that since the columns of the matrix X are linearly dependent, owing to temporal correlation among the frames, it is possible to recover X by low-rank recovery techniques (10.57) as follows,

$$\hat{X} = \min_X \left\| X \right\|_* \text{ subject to } \left\| Y - RFX \right\|_2^2 \leq \varepsilon \tag{10.59}$$

It is known that CS recovery can yield good reconstruction results where as low-rank recovery is more of a proof-of-concept. The natural question that arises is: can the two be combined to yield even better results? The first study that exploited this possibility is called k-t sparse and low rank (SLR). They proposed to recover the dynamic MRI sequence by posing it as the following unconstrained reconstruction problem,

$$\hat{X} = \min_{X} \left\| Y - RFX \right\|_{Fro}^{2} + \lambda_1 \left\| W \otimes F_{1D}(X) \right\|_1 + \lambda_2 \left\| X \right\|_{S_p} \tag{10.60}$$

It must be noted that the Schatten-p ($0 < p < 1$) is used as a low-rank penalty.

Experimental Evaluation

Experimental results are reported on a physiologically improved NCAT (PINCAT) phantom. An acceleration factor of 5 is used for the simulations. The reconstructed and difference images are shown in Figure 10.21. It is easy to observe that results from low-rank reconstruction are the worst. There is a lot of error along the edges. CS reconstruction improves over low-rank reconstruction but is still not the best. The best result is obtained by combining CS with low-rank recovery. For all the experiments, the value of p (of Schatten-p norm) is fixed at 0.1.

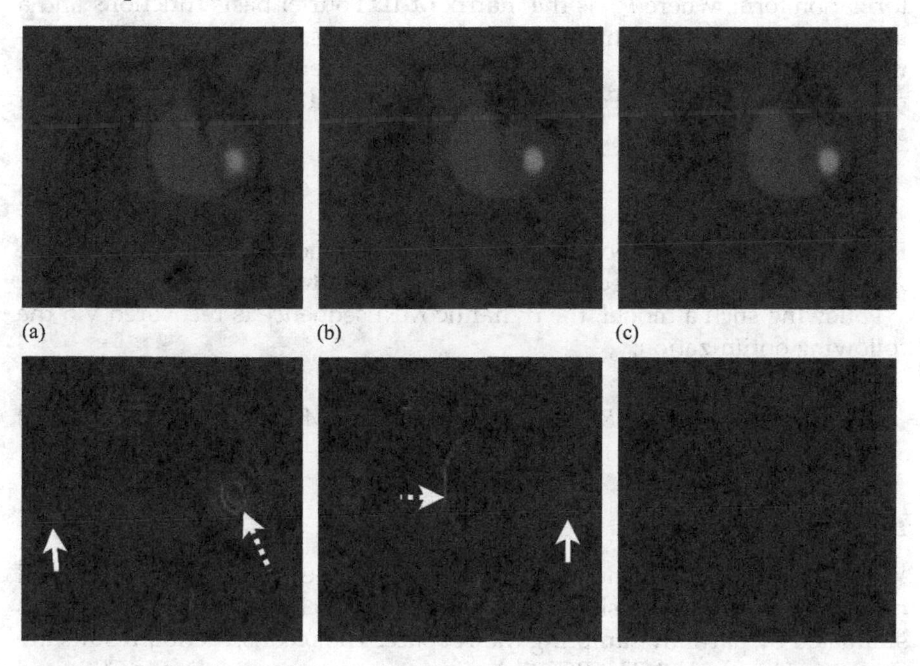

FIGURE 10.21
(a) Low-rank reconstruction, (b) CS reconstruction, and (c) combined low-rank and CS reconstruction. Top row: reconstructed images, and bottom row: difference images.

Blind Compressive Sensing

This is one of the most recent techniques in dynamic MRI reconstruction. This method is neither a pure CS-based method nor a rank-recovery-based technique; yet, it contains flavors of both. Thus, we have decided to discuss it under a separate sub-section.

The columns of the dynamic MRI matrix are linearly dependent; this is because the frames are temporally correlated. Therefore, it is possible to express the dynamic MRI matrix as a linear combination of a few temporal basis functions. The concept follows from previous studies where it is assumed X is sparse in x-f space. Sparsity in x-f space can be expressed as follows,

$$XF_{1D}^{T} = \Upsilon \text{ (sparse coefficient matrix)} \tag{10.61}$$

Alternately,

$$X = \Upsilon F_{1D} \tag{10.62}$$

In expression (10.62), the dynamic MRI matrix X is expressed in a matrix factorization form, where F_{ID} is the matrix of 1D Fourier basis functions and a matrix of sparse coefficients Υ. In the blind compressed sensing (BCS) framework for dynamic MRI reconstruction, instead of employing a fixed temporal basis such as the 1D Fourier transform (10.62), the basis is learnt during signal reconstruction,

$$X = \Upsilon M \tag{10.63}$$

Here, M is a learned temporal basis that sparsifies the dynamic MRI sequence X along the temporal direction. This matrix needs to be learnt from the data.

Following such a model, the dynamic MRI sequence is recovered via the following optimization,

$$\hat{X} = \min_{\Upsilon, M} \|Y - RF(\Upsilon, M)\|_{Fro}^{2} + \lambda_1 \|\Upsilon\|_1 + \lambda_2 \|M\|_{Fro}^{2}, X = \Upsilon M \tag{10.64}$$

Experimental Evaluation

We report the results from the published work. The experiments were carried out on myocardial perfusion MRI dataset. An acceleration factor of 7.5 was simulated by partially sampling the K-space. The reconstruction results are shown in Figure 10.22. The BCS technique was compared against a CS-based method that assumed sparsity in the transform domain and also with the

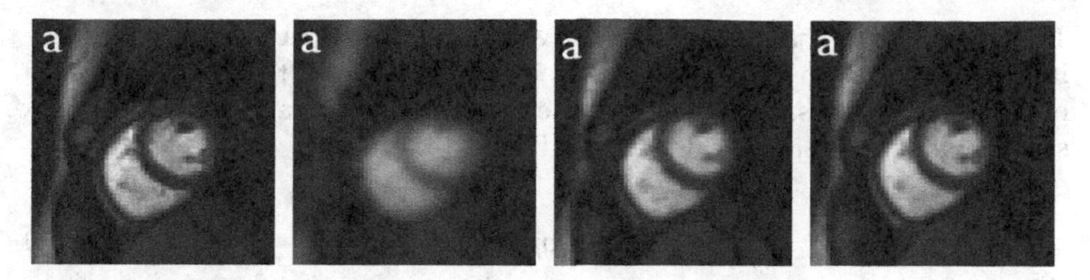

FIGURE 10.22
Left to right: ground-truth, zero-filling, compressed sensing (assuming sparsity in x-f space), and BCS.

naïve zero-filling technique, where the unsampled K-space locations were filled with zeroes and the image was reconstructed by inverse FFT. The quality of reconstruction is easily discernible from the reconstructed images, and hence, we do not provide the difference images. The learned basis in BCS yields better reconstruction than the fixed Fourier basis (CS).

Sparse + Low-Rank Reconstruction

In robust principal component analysis (RPCA), the problem is to decompose a signal into its sparse and low-rank components. This was successfully used for background-foreground separation in videos. Let x^t be the tth video frame. It can be assumed to be composed of two parts: a slow-moving background and a fast-moving foreground. Also, the foreground typically comprises a much smaller region compared with the background.

Each frame is assumed to be formed by a super-position of background (b) and foreground (f), that is, $x_t = b_t + f_t$. Assume that a Casorati matrix is formed by stacking all the frames as columns of a matrix: $X = [x_1]\ldots[x_T] = [b_1]\ldots[b_T] + [f_1]\ldots[f_T]$. The background changes very slowly with time; therefore, the Casorati matrix corresponding to the background is of very low rank, owing to temporal correlations among the backgrounds in different frames; let $L = [b_1|\ldots|b_T]$. The foreground occupies a small area of the entire frame, and in turn, after background subtraction, the foreground is sparse. Therefore, the matrix $S = [f_1|\ldots|f_T]$ will be sparse. Thus, the Casorati matrix X can be represented as,

$$X = S + L \tag{10.65}$$

RPCA techniques can be employed to separate the two components by solving the following optimization problem,

$$\min_{S,L} \|S\|_1 + \lambda \|L\|_* \text{ subject to } X = S + L \tag{10.66}$$

The problem in dynamic MRI is more complicated, but RPCA could be modified for this purpose. The major change from RPCA is that in MRI, we do not have the fully sampled data; rather, we only have partial K-space measurements. Therefore, (10.66) needs to be modified as,

$$\min_{S,L} \|S\|_1 + \lambda \|L\|_* \text{ subject to } Y = F(S+L) \tag{10.67}$$

where Y is the Casorati matrix formed by stacking the K-space measurements of each frame as columns and F is the Fourier mapping from the image domain to the K-space.

This formulation is good for functional magnetic resonance imaging (fMRI) problems where the active foreground occupies a very small area. However, this formulation cannot be generalized. Typically, in dynamic MRI, for example, cardiac cine MRI or cardiac perfusion MRI, the active foreground is large; in such a case, it is not judicious to assume the foreground to be sparse. To overcome this problem, transform domain sparsity is incorporated to improve upon it; that is, the foreground is assumed to be sparse in a transform domain. The recovery is modified as follows,

$$\min_{S,L} \|TS\|_1 + \lambda \|L\|_* \text{ subject to } Y = F(S+L) \tag{10.68}$$

Here, T is the sparsity-promoting transform (Figure 10.23).

In practice, the equality constraint is relaxed, and the following problem is solved instead,

$$\min_{S,L} \|Y - F(S+L)\|_F^2 + \lambda_1 \|TS\|_1 + \lambda_2 \|L\|_* \tag{10.69}$$

We have already learnt in Chapter 1 how to solve this problem.

Experimental Evaluation

2D cardiac cine imaging was performed in a healthy adult volunteer by using a 3T scanner (Tim Trio, Siemens Healthcare, Erlangen, Germany). Fully sampled data were acquired using a 256 × 256 matrix size (FOV = 320 × 320 mm^2) and 24 temporal frames and were retrospectively undersampled by factors of 4, 6, and 8 by using a k-t variable-density random undersampling scheme.

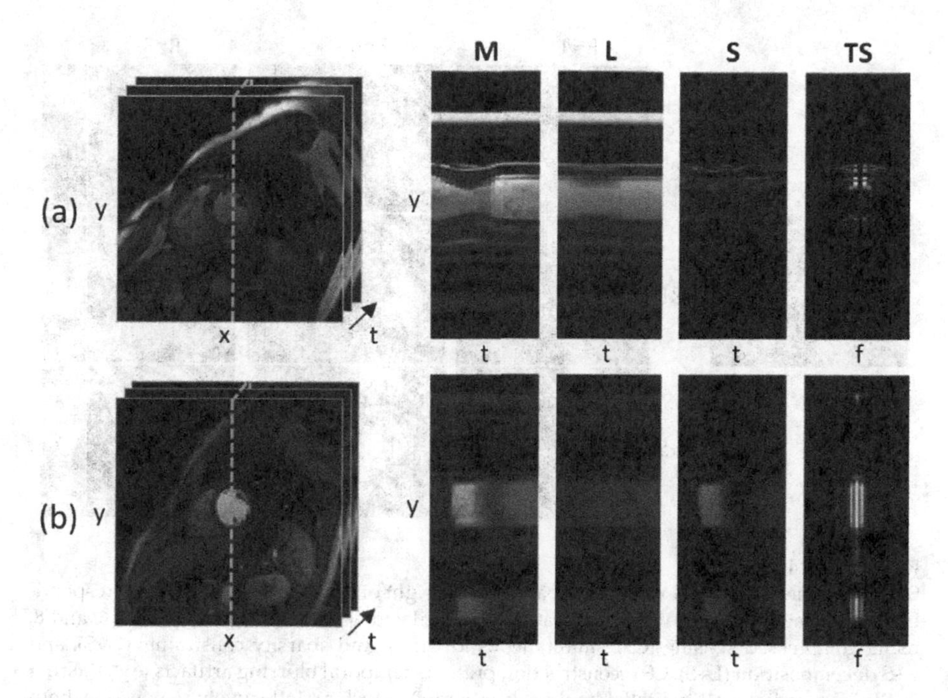

FIGURE 10.23
L+S decomposition of fully sampled 2D cardiac cine (a) and perfusion (b) datasets corresponding to the central x location. The low-rank component L captures the correlated background, and the sparse component S captures the remaining dynamic information. The rightmost column shows the sparse component S in transform domain (Fourier transform along temporal direction), which shows increased sparsity compared with the original y-t domain.

Image reconstruction was performed using multicoil CS, k-t SLR, and sparse plus low-rank (L+S) methods, with a temporal Fourier transform serving as sparsifying transform. Quantitative image quality assessment was performed using root mean squared error (RMSE) and structural similarity index (SSIM) metrics, as described in the cardiac perfusion example (Figure 10.24).

The results were expected. As is evident from Table 10.2, the L+S approach yields lower RMSE and higher SSIM than both CS and k-t SLR. CS introduced temporal blurring artifacts, particularly at systolic phases, where the heart is contracted and the myocardial wall is prone to signal leakage from other frames. Both k-t SLR and L+S reconstructions can significantly remove these artifacts, but the L+S reconstruction offers an improved estimation of the original cine image, as depicted by better preservation of fine structures in the x-t plots and reduced residual aliasing artifacts. This fact is due to

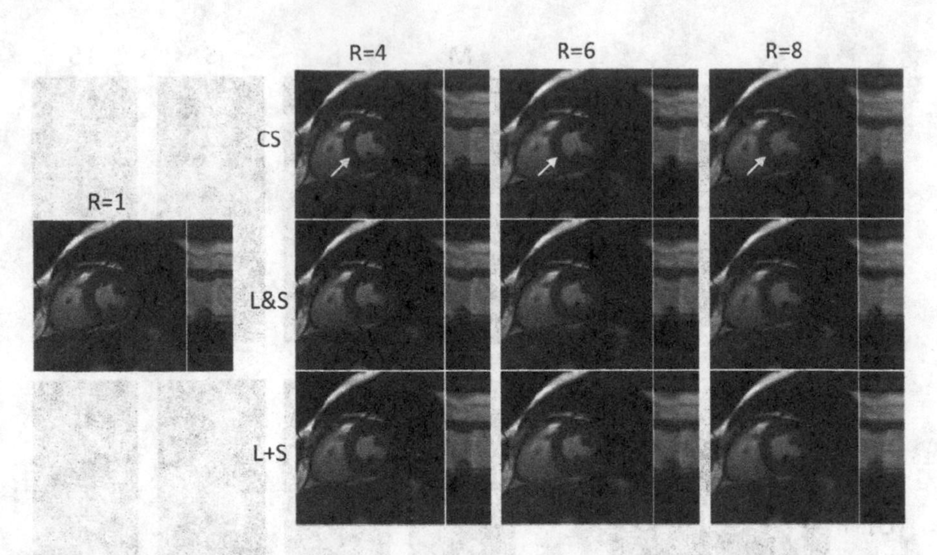

FIGURE 10.24
Systolic phase images and x-t plots (in panels to the right of the short-axis images) corresponding to reconstruction of cardiac cine data with simulated acceleration factors R = 4, 6, and 8, using compressed sensing (CS), simultaneous low-rank and sparsity constraints (L&S), and L+S decomposition (L+S). CS reconstruction presents temporal blurring artifacts (e.g., the ring in the myocardial wall indicated by the white arrow), which are effectively removed by both L&S and L+S reconstructions. However, L+S presents higher temporal fidelity (fine structures in the x-t plots) and lower residual aliasing artifacts.

TABLE 10.2
RMSE and SSIM Values

Method	Accn Factor = 4	Accn Factor = 6	Accn Factor = 8
CS	6.91/0.966	10.01/0.944	13.38/0.918
k-t SLR	7.22/0.958	9.42/0.946	11.91/0.928
L+R	**6.23/0.970**	**8.04/0.956**	**9.69/0.934**

Accn = acceleration

the background suppression performed by the L+S reconstruction, which provides a sparser S and thus facilitates accurate reconstruction of undersampled data. The background estimated in the L component is not stationary over time and contains the most correlated motion. The S component contains the cardiac motion with larger variability.

Suggested Reading

J. Song, Q. H. Liu, G. A. Johnson, and C. T. Badea, Sparseness prior based iterative image reconstruction for retrospectively gated cardiac micro-ct, *Medical Physics*, 34, 4476–4482, 2007.

G. H. Chen, J. Tang and S. Leng, Prior image constrained compressed sensing (PICCS): A method to accurately reconstruct dynamic CT images from highly undersampled projection data sets, *Medical Physics*, 35 (2), 660–663, 2008.

M. Lustig, D. L. Donoho and J. M. Pauly, Sparse MRI: The application of compressed sensing for rapid MR Imaging, *Magnetic Resonance in Medicine*, 58 (6), pp. 1182–1195, 2007.

K. P. Pruessmann, M. Weiger, M. B. Scheidegger and P. Boesiger, SENSE: Sensitivity encoding for fast MRI, *Magnetic Resonance in Medicine*, 42, 952–962, 1999.

M. Blaimer, F. Breuer, M. Mueller, R. M. Heidemann, M. A. Griswold and P. M. Jakob, SMASH, SENSE, PILS, GRAPPA: How to choose the optimal method, *Review article in Topics in Magnetic Resonance Imaging*, 15 (4), 223–236, 2004.

D. Liang, B. Liu, J. Wang and L. Ying, Accelerating SENSE using compressed sensing, *Magnetic Resonance in Medicine*, 62 (6), 1574–1584, 2009.

L. Ying and J. Sheng, Joint image reconstruction and sensitivity estimation in SENSE (JSENSE), *Magnetic Resonance in Medicine*, 57, 1196–1202, 2007.

A. Majumdar and R. K. Ward, Calibration-less multi-coil MR image reconstruction, *Magnetic Resonance Imaging*, 30 (7), 1032–1045, 2012.

A. Majumdar and R. K. Ward, Accelerating multi-echo T_2 weighted MR imaging: Analysis prior group sparse optimization, *Journal of Magnetic Resonance*, 210 (1), 90–97, 2011.

S. G. Lingala, Y. Hu, E. DiBella, and M. Jacob, Accelerated dynamic MRI exploiting sparsity and low-rank structure, *IEEE Transaction Medical Imaging*, 30, 1042–1054, 2011.

S. G. Lingala and M. Jacob, Blind compressed sensing dynamic MRI, *IEEE Transactions on Medical Imaging*, 32, 1132–1145, 2013.

11

Biomedical Signal Reconstruction

Wireless telemonitoring of physiological signals provides an opportunity for compressed sensing (CS). Imagine a scenario where the subject wears some sensor nodes that acquire his/her biomedical signals, and these signals are transmitted instantaneously to a remote facility, where the signals are monitored and analyzed. However, current systems for acquiring some common signals, such as electroencephalogram (EEG)/electrocardiogram (ECG)/pulse pleithismogram (PPG)/magnetoencephalogram (MEG), are wired, cumbersome, and need trained personnel to use them. One cannot expect the elderly population to undergo such training. For them, the acquisition devices must be simple, portable, and wireless.

We envisage a system whereby wireless sensor nodes are used to collect the data and transmit them wirelessly. As the sensor nodes do not have the computational capability to compress the signals, the sampled signals are transmitted to a local smartphone (and not directly to a remote destination). This is because the transmission to a remote location wirelessly would consume a lot of power. The smartphone would process the collected data, compress them, and transmit them to the remote medical facility for further analysis. Today's smartphones are packed with powerful processors and can easily carry out the aforesaid tasks.

The main constraint in such a wireless body area network (WBAN) is the energy consumption at the sensor nodes. Since the desired device is a portable wireless system, the battery would be small. Therefore, the sensor should be designed such that the battery lasts as long as possible. There are three power sinks in a WBAN: sensing, processing, and transmission. The energy required for transmission is the highest; therefore, every effort needs to be made to reduce the transmission cost. One can save the transmission energy by compressing the signal. However, modern compression techniques require considerable processing power; the portable sensor nodes are not endowed with such capabilities, and hence, such compression techniques cannot be implemented directly at the sensor nodes.

To overcome this issue, prior studies have proposed CS-based compression techniques. Here, the signal (at the sensor node) is first compressed by projecting it onto a lower dimension by a random projection matrix. This is a simple operation (matrix vector product) and can be implemented on a digital signal processing (DSP) chip (or even hard-wired for the sake of efficiency). The resulting compressed signal is then transmitted (to a local smartphone), where the original signal is recovered from the compressed one via CS recovery techniques. The assumption is that the computational power at the receiving point (in our case the smartphone) is not a bottleneck.

Sparse Recovery

As mentioned earlier, the simplest solution for the problem is to project the acquired signal (x) from a single channel onto a random matrix (M) in order to obtain a compressed version (y). This is expressed as,

$$y = Mx \tag{11.1}$$

Biomedical signals are known to be sparse in some known basis, such as wavelet, discrete cosine transform (DCT), and Gabor. For orthogonal or tight-framed basis (S), one can use the synthesis formulation to express (11.1) as follows,

$$y = MS^T\alpha \tag{11.2}$$

Here, S is the forward transform and α represents the sparse coefficients. Therefore, one can solve (11.2) by using any standard CS recovery algorithm such as l_1 minimization (say).

$$\min_{\alpha} \|\alpha\|_1 \text{ subject to } y = MS^T\alpha \tag{11.3}$$

Once the sparse coefficient α is obtained from (11.3), one applies the inverse transform to get the recovered signal ($x = S^T\alpha$).

However, as is known, the synthesis prior restricts the choice of sparsifying transforms; for example, one cannot use Gabor since it is not orthogonal or tight-frame in general. In order to overcome this restriction, one can employ the analysis prior formulation. As we learnt before, this can be applied to any linear basis and hence is more generic. Assuming that the sparsity basis is S, the analysis formulation can be expressed as,

$$\min_{x} \|Sx\|_1 \text{ subject to } y = Mx$$

The analysis prior directly solves for the signal.

Structured Sparse Recovery of Multi-channel Signals

Most biomedical signals are acquired via multiple channels, for example, EEG/ECG. One can definitely apply the aforesaid solution to each of the channels and get back the reconstructed signal. But this is not optimal, since it will not capture the inherent structure of the multi-channel data.

For better understanding, consider the EEG signal captured by four channels in Figure 11.1a. The underlying brain activity that is evoking the signal is unique. It is just that, owing to the difference in position of the electrodes, each channel is capturing a slightly different (filtered) version. This is the reason that the basic pattern remains the same in all the channels. In order to optimally reconstruct the multi-channel signals, such structural information needs to be used.

It is well known that wavelet (and most other sparsifying transforms) captures the singularities in the signal. Since all the channels show the same pattern, the positions of the singularities do not change (even though their values do). Hence, when a sparsifying transform is applied on the multi-channel signals, the positions of the non-zero values remain the same across the channels—this is evident from Figure 11.1b. Therefore, if the wavelet coefficients of different channels are stacked as columns of a matrix, the corresponding matrix will have a row-sparse structure. Only those rows corresponding to the singularities in the original signal will have non-zero values.

The multi-channel acquisition problem can be modeled as follows,

$$y_i = Mx_i = MS^T\alpha_i, \text{ for ith channel} \tag{11.4}$$

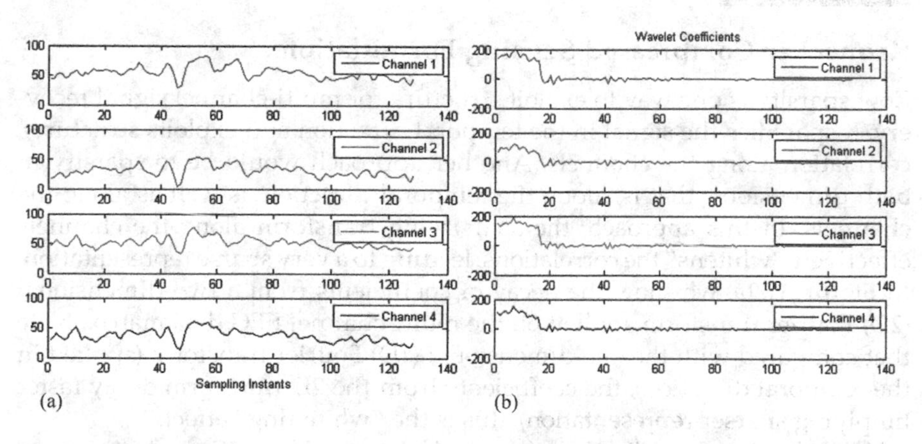

FIGURE 11.1
(a) EEG signal acquired by four channels. (b) Wavelet transform of corresponding channels.

Expressing (11.4) in terms of all channels,

$$Y = MX = MS^T A \tag{11.5}$$

Here Y, X, and A are formed by stacking the y_is, x_is and α_is as columns, respectively.

We have argued that the matrix of coefficients (A) will be row-sparse. This is the underlying structural information that one can exploit for the said problem. We have learnt that such row-sparse multiple measurement vector recovery can be posed as $l_{2,1}$ minimization of the following form,

$$\min_A \|A\|_{2,1} \text{ subject to } Y = MS^T A \tag{11.6}$$

One should recall the definition of $l_{2,1}$-norm; it is the sum of the l_2-norms of the rows. The inner l_2-norm ensures that the selected rows have dense values, while the outer l_1-norm promotes sparsity in the selection of rows.

As before, one needs to apply the inverse of the sparsifying transform on the recovered coefficients to get the signal.

The aforesaid solution (11.6) is the synthesis prior formulation and is applicable only to orthogonal basis or tight-frame transforms. A more generic approach to solve the multi-channel recovery problem would be to pose it as an analysis prior form.

$$\min_X \|SX\|_{2,1} \text{ subject to } Y = MX \tag{11.7}$$

Kronecker Compressed Sensing Formulation

Row sparsity is one way to exploit structure for multi-channel signal recovery. It sparsifies the signal in the temporal direction and exploits structure/correlation along the channels. Another approach would be to sparsify in both dimensions, that is, along the temporal direction as well as along the channels. In this approach, the sparsifying transform along the channels effectively "whitens" the correlations, leading to a very sparse representation. In Figure 11.2a, we show the decay of coefficients from a two-dimensional (2D) Fourier transform applied on the multi-channel EEG data matrix. Note that compared with the one-dimensional (1D) Fourier transform (applied in the temporal direction), the coefficients from the 2D transform decay faster (implying sparser representation); this is the "whitening" effect.

If X is the signal, application of a sparsifying transform (S_1) in the temporal direction would be represented as S_1X, and application of some transform

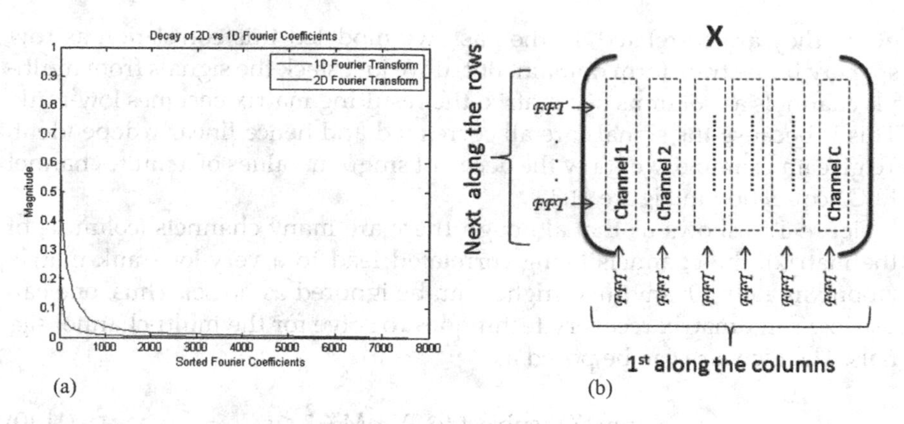

(a) (b)

FIGURE 11.2
(a) Decay of coefficients. (b) Schematic diagram for application of 2D transform.

(S_2) along the channels will be expressed as XS_2. The combined application of the two transforms will be expressed as S_1XS_2. The schematic diagram is shown in Figure 11.2b.

Let us consider that the sparse coefficients after application of the 2D transform is Z, that is, $Z = S_1XS_2$; we can express the multi-channel data matrix X as $S_1^TZS_2^T = X$, assuming that the transforms are orthogonal or tight-frame. Plugging this expression into the multi-channel acquisition model (5) leads to,

$$Y = MX = MS_1^TZS_2^T \tag{11.8}$$

In the Kronecker notation, this can be alternately expressed as,

$$vec(Y) = MS_2 \otimes S_1^T vec(Z) \tag{11.9}$$

The sparse matrix Z can be recovered by l_1 minimization or via some greedy technique. Once Z is recovered, one can obtain the multi-channel signal matrix by applying $S_1^TZS_2^T = X$. This is the so-called Kronecker CS formulation.

Low-Rank Recovery

Yet another way to recover multi-channel biomedical signals is to employ matrix recovery techniques. We would request the reader to divert their attention to Figure 11.1a once more. Since all the channels sample the same event, they are correlated. In the past, we modeled this correlation as row sparsity in the transform domain. But, if we just stack the signals from multiple channels as columns of a matrix, the resulting matrix becomes low-rank. This is because the signals are all correlated and hence linearly dependent. To give an example, we show the decay of singular values of a multi-channel EEG acquisition in Figure 11.3.

Figure 11.3 shows us that although there are many channels (columns of the matrix), the channels being correlated lead to a very-low-rank matrix (approximately 10; anything higher can be ignored as noise). Thus, one can use low-rank matrix recovery techniques to solve for the multi-channel signals. The recovery can be posed as,

$$\min_X \|X\|_* \text{ subject to } \|Y - MX\|_F^2 \leq \varepsilon \tag{11.10}$$

Note that for this solution, having the inequality constraint is mandatory. This is because the model is not exactly low-rank but approximately so. Therefore, we had to model the higher singular values as noise.

One must remember that this formulation is quite restrictive. It cannot be applied to matrices that are too tall (too few channels) or too fat (too many channels, which is not a practical scenario). The method cannot be applied when the number of channels is limited; that is, it cannot be applied for most ECG data, since one usually gathers it with few leads. It cannot be applied for EEG data from most low-cost acquisition devices, since the number of channels

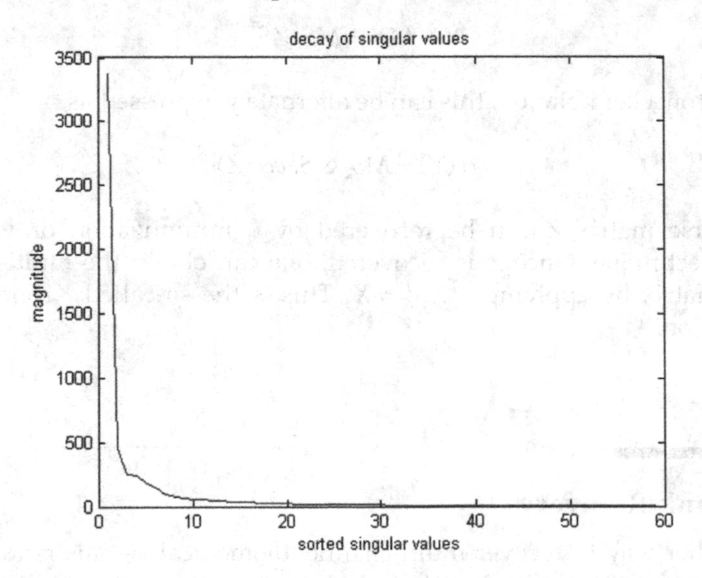

FIGURE 11.3
Decay of singular values.

rarely goes beyond 20. But, for medical-grade EEG, when the number of channels is 32 or more, such low-rank recovery techniques can be applied.

Blind Compressed Sensing

When the situation is conducive for matrix recovery, it is recommended to use blind CS. This is because the method is non-parametric, and one does not need to figure out what is the best possible sparsifying basis for the given signal.

The nuclear norm-based minimization technique for low-rank matrix recovery is relatively new. Previous studies used to employ matrix factorization for the same. They factored the low-rank matrix X in terms of a skinny matrix U and a tall matrix V. Instead of recovering X directly, they solved for U and V by posing it as a matrix factorization problem. When applied to the multi-channel reconstruction problem, this can be expressed as follows,

$$\min_{U,V} \|Y - MUV\|_F^2 \tag{11.11}$$

The matrices U and V are updated via alternate minimization till the cost function reaches a local minimum. The problem (11.11) is non-convex but is simple and efficient.

Such factorization-based techniques can also be interpreted through CS. Instead of assuming the sparsifying basis to be known (DCT, wavelet, etc.), one can learn it from the data (provided there are enough samples). If X is the data, one can learn a basis D and the corresponding coefficients in the same manner as matrix factorization. Since CS requires the coefficients to be sparse, there is an additional sparsity constraint on the learnt coefficients.

$$\min_{D,Z} \|Y - MDZ\|_F^2 + \lambda \|Z\|_1 \tag{11.12}$$

Usually, there is a constraint on D to prevent degenerate solutions where D is very low and Z is very high or vice versa. The easiest way to prevent this is to impose unit norm on the columns of D.

The said formulation is called blind CS. It is blind because the sparsifying basis is not known and is learnt from the data. One can add sophistication to the basic formulation (11.12). Instead of learning the basis only in the temporal direction, one can learn it in both dimensions. This leads to,

$$\min_{D_1,D_2,Z} \|Y - MD_1ZD_2\|_F^2 + \lambda \|Z\|_1 \tag{11.13}$$

This is akin to the Kronecker formulation. Instead of sparsifying in both directions, once can exploit the structure of the multi-channel signal via row sparsity.

$$\min_{D,Z} \|Y - MDZ\|_F^2 + \lambda \|Z\|_{2,1} \tag{11.14}$$

Results

Here, we will show some reconstruction results. These are on EEG reconstruction where the original signal has been undersampled by 50% and recovered by using various reconstruction techniques (Figure 11.4).

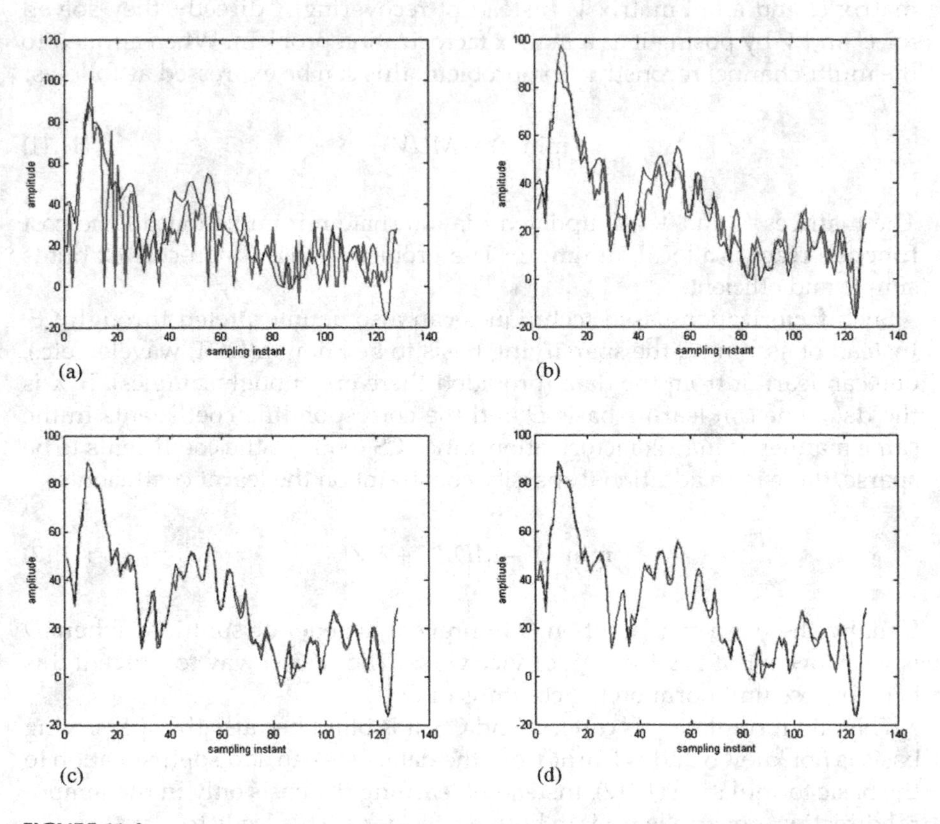

FIGURE 11.4
Blue—actual signal and red—reconstructed signal. (a) Synthesis prior CS reconstruction. (b) Analysis prior CS reconstruction. (c) Kronecker CS reconstruction. (d) Row-sparse reconstruction.

(*Continued*)

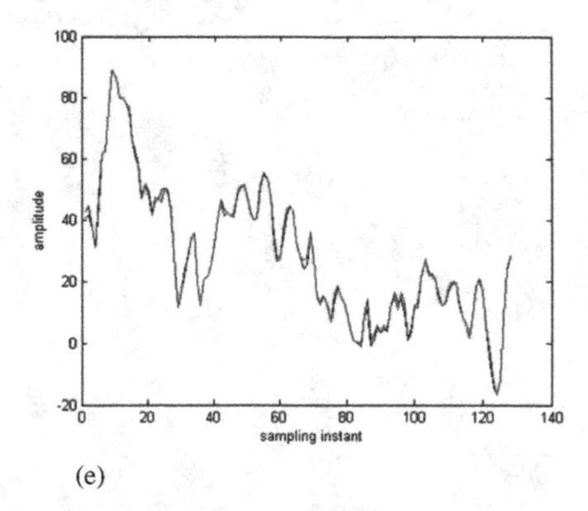

(e)

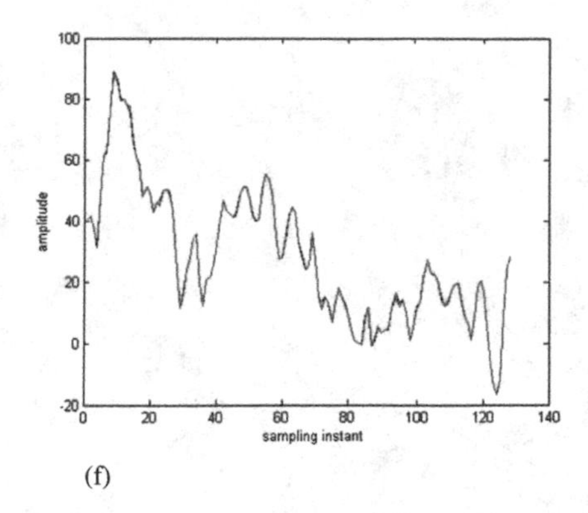

(f)

FIGURE 11.4 (Continued)
Blue—actual signal and red—reconstructed signal. (e) Kronecker blind CS reconstruction. (f) Row-sparse blind CS reconstruction.

The results show that the synthesis prior reconstruction is the worst. It does not reconstruct the signal even remotely. The analysis prior formulation is slightly better; at least, it follows the general pattern of the signal. The Kronecker CS and the row-sparse reconstructions are better; there is only small deviations from the original signal. But the best results are obtained from the blind CS reconstruction techniques.

Suggested Reading

S. Aviyente, Compressed sensing framework for EEG compression, *IEEE Workshop on Statistical Signal Processing*, pp. 181, 184, 2007.

Z. Zhang, T. P. Jung, S. Makeig and B. D. Rao, Compressed sensing of EEG for wireless telemonitoring with low energy consumption and inexpensive hardware, *IEEE Transactions on Biomedical Engineering*, 60 (1), 221–224, 2013.

M. Mohsina and A. Majumdar, Gabor based analysis prior formulation for EEG signal reconstruction, *Biomedical Signal Processing and Control*, 8 (6), 951–955, 2013.

A. Shukla and A. Majumdar, Exploiting inter-channel correlation in EEG signal reconstruction, *Biomedical Signal Processing and Control*, 18 (4), 49–55, 2015.

A. Majumdar and R. K. Ward, Energy efficient EEG sensing and transmission for wireless body area networks: A blind compressed sensing approach, *Biomedical Signal Processing and Control*, 20, 1–9, 2015.

12

Regression

Least Angle Selection and Shrinkage Operator

Consider a hypothetical problem of infant mortality in developing countries; this is a big issue, since the mortality rates are high. When a pregnant mother comes to a hospital, she needs to fill out a questionnaire. Typically, state-run hospitals in India collect information such as age of the mother, age of the father, their education, their income, their food habits (non-vegetarian/vegetarian/vegan), the number of times they visit the doctor, and their location. Since infant mortality is a serious issue, doctors try to apply regression techniques to zero in factors that might be responsible for the outcome.

This can be expressed as a regression problem. The factors (data collected from the questionnaire) that are supposed to be responsible are the explanatory variables along the columns of the vector a_i and the outcome is y_i (mortality). In linear regression, it is assumed that the output is produced as a linear combination (w_i) of the explanatory variables. This is expressed as,

$$y_i = a_i^T w_i \tag{12.1}$$

When the data is expressed for all patients, (12.1) can be written as,

$$y = Aw + n \tag{12.2}$$

Here, y is the vector of all the outcomes, A is the matrix of explanatory variables (a_is) arranged along the rows, and w is the linear weight vector. Since such regression models can never be perfect, it is common to assume the presence of some Gaussian noise n.

The most straightforward approach to solve this problem is via least squares minimization, that is,

$$w = \min_w \|y - Aw\|_2^2 \tag{12.3}$$

Usually, the solution is regularized via a Tikhonov penalty, leading to the ridge regression. This is expressed as,

$$w = \min_{w} \|y - Aw\|_2^2 + \lambda \|w\|_2^2 \tag{12.4}$$

We know that such a solution would produce dense values in w, and hence, the solution in itself would not be very helpful. Because, in this case, the goal is not so much as to predict the outcome of the new mother, given her inputs, but to analyze the reasons for infant mortality from past data. In such a case, a dense w would mean that all the factors are responsible for the outcome (mortality). It is not possible to control all the factors; for example, it is not possible to control the ages of mother and father or their location, but the dense solution would say that one needs to control all of them in order to check infant mortality. Such a solution would indeed be fairly useless.

Please remember that we are not saying that ridge regression is useless. It is just that in this scenario or other such scenarios such as gene expression analysis,[1] where the objective is to select a sparse set of factors so as to control them, ridge regression would be useless. However, following traditional statistical techniques such as analysis of variance (ANOVA) coupled to ridge regression have proven to yield interesting results in the past.

When the task is to select a sparse set of factors, in order to control them, one needs to employ the l_1-norm penalty instead of the Tikhonov regularization. In signal processing, this was known as basis pursuit denoising (BPDN); in machine learning, this has been known as Least Angle Selection and Shrinkage Operator (LASSO), a term introduced in Tibshirani's seminal paper.[2] The original paper proposed solving the sparse factor selection problem as,

$$\min_{w} \|y - Aw\|_2^2 \text{ such that } \|w\|_1 \leq \tau \tag{12.5}$$

Note the slight deviation in formulation from the standard signal processing literature. Instead of applying the constraint on the data fidelity, it is applied on the sparsity. This is because, in signal processing, it is usually fairly simple to estimate the noise level, and hence, a constraint on the data fidelity is sensible. But in machine learning, in problems where the data is not generated by electrical systems, estimating the noise level is impossible. It is more

[1] J. Lucas, C. Carvalho, Q. Wang, A. Bild, J. R. Nevins and M. West, Sparse statistical modelling in gene expression genomics. *Bayesian Inference for Gene Expression and Proteomics*, 1, 0–1, 2006.

[2] R. Tibshirani, Regression shrinkage and selection via the lasso. *Journal of the Royal Statistical Society. Series B (Methodological)*, 267–288, 1996.

sensible to constrain the number of sparse factors. That being said, it is not easy to solve the LASSO problem directly; usually, its simpler unconstrained version is solved instead.

$$\min_{w} \|y - Aw\|_2^2 + \lambda \|w\|_1 \tag{12.6}$$

By varying the value of λ, one can get either a very sparse (high-value) or a dense (low-value) solution.

Elastic Net

LASSO is good when the variables are independent—it can select them. But when they are not independent, that is, when the factors/variables are correlated, LASSO fails to pick all of them. It will only select the variable that is most correlated with the output. Consider the repercussion of our infant mortality example. The hospital would have collected the information about income level, their food habits, and their education level. Intuitively, we can understand that all the factors would be correlated and would affect the said outcome. However, if we run LASSO on the data, it is highly likely that we will end up with only one variable and that variable will be income level or education level, because these variables would be most correlated with the output. But such an analysis would not be very useful. The doctors cannot control the parents' income or education.

In the aforesaid scenario, it would be useful if there was a mechanism that would detect all the three factors—income, nutrition, and education. In that case, the doctors could probably advise the parents to change the mother's food intake for the period of pregnancy in order to improve her chances. But clearly, LASSO would not be a solution for that; it is likely that it would completely miss on selecting nutrition as a factor. To ameliorate this issue, elastic net was proposed. The expression of elastic net is:

$$\min_{w} \|y - Aw\|_2^2 + \lambda_1 \|w\|_1 + \lambda_2 \|w\|_2^2 \tag{12.7}$$

Let us analyze each of the terms. The first one is the standard data-fidelity term. The second one is a LASSO-like term, which promotes sparsity in the selection of the variables. The third term is a Tikhonov-like penalty. It promotes selection of all the factors. The interplay between the second and third terms leads to the "elasticity" of the solution. The sparsity promoting l_1-norm tries to select very few factors; on the other hand, the l_2-norm attempts to select all of them. The combined effect of the two terms leads to the selection of correlated factors.

Coming back to our illustrative problem, even though LASSO would have missed on picking up nutrition as a factor, elastic net is likely to have picked it up (along with education and income). In such a case, the doctors would have discarded the factors that they could not control but would have advised the mother to take proper diet.

Group LASSO

LASSO is good for selecting independent variables that are responsible for an output, but it fails to capture a set of correlated variables. This issue is overcome by the elastic-net formulation. It is a good option when the variables are known to be correlated but the exact group of correlated variables is unknown. The degree of sparsity and density is controlled by the relative values of λ_1 and λ_2 (12.7). However, unlike the LASSO formulation, where some estimate is there on the sparsity level, there is simply no way that one could know the degree of correlation between the variables and select the values of the two parameters (11.7) accordingly. Therefore, elastic net, although sound in practice and useful when no other information is available, is a relatively gray area as far as parameter tuning is concerned.

When the group of correlated variables is known, one can go for a supervised regression approach, where all the group of variables are selected. Unfortunately, it is not possible to give a layman's examples (as infant mortality) to explain this problem. We take a problem from bioinformatics.[3] Assume that we are studying cancer at the genetic level. For a particular type of cancer, we have to find out the genes that are responsible. One can apply LASSO and get back results that are not all encompassing or can apply elastic net without having any idea what the results mean (since one cannot select the parametric values), but a better approach would be to select gene clusters where the membership of the genes to different clusters is already available (from other sources). This would guarantee selection of all the important groups of genes responsible for the type of cancer we are studying.

Suppose that the genes are already grouped into n clusters and that the output is dependent on all the genes. This can be expressed as,

[3] S. Ma, X. Song and J. Huang, Supervised group Lasso with applications to microarray data analysis. *BMC Bioinformatics*, 8 (1), 60, 2007.

$$y_i = [a_{j1}^T, a_{j2}^T, ..., a_{jn}^T] \begin{bmatrix} w_1 \\ w_2 \\ ... \\ w_n \end{bmatrix} \quad (12.8)$$

Here, a_{jr} means that these are the genes for the rth cluster responsible for the ith output; similarly, w_r represents the weights for the rth cluster.

If we consider all the samples, we can represent (12.8) as follows,

$$y = [a_1, a_2, ..., a_n] \begin{bmatrix} w_1 \\ w_2 \\ ... \\ w_n \end{bmatrix} \equiv y = Aw \quad (12.9)$$

Here, a_1, a_2, and a_n represent the matrices formed by genes for all the samples and y is the output stacked vertically as rows. Our goal is to find out the genetic clusters that are responsible for the output. Since the cluster information is available, one can recast it as a group sparsity problem. We have already studied the formulation and the algorithms for solving such problems before. We know that the solution can be expressed as follows,

$$\min_w \|y - Aw\|_2^2 + \lambda \|w\|_{2,1} \quad (12.10)$$

The structural information is embedded in the $l_{2,1}$-norm. It is defined as the sum of the l_2-norms of the groups/clusters. We know that such mixed-norm minimization problems lead to group sparse solutions. The l_1-norm enforces selection of very few groups, and the l_2-norm enforces selection of all genes in the chosen groups.

Multi-task Regression

So far, we have been discussing problems where the objective is to explain a single type of output variable—infant mortality, cancer, etc. In multi-task regression, the goal is to explain a variety of outputs.

Let us suppose that we have to analyze data from a group of freshman students. We have their marks for different subjects from their examinations (say social sciences, mathematics, and sciences), and we have information captured during their admission process, for example, their score in scholastic aptitude test (SAT), their high school results in different subjects, parents' education level, and location of their high school. The task is to find out what are the predominant existing factors that are responsible for their examination results. Note that in this case, we have multiple outputs—social sciences, mathematics, and sciences and a common set of input variables. Had we framed a typical (single-task) regression problem, we would have expressed it as,

$$\min_{w_i} \|y_i - Aw_i\|_2^2 + \lambda \|w_i\|_1 \tag{12.11}$$

Here, i denotes the subject (task). By solving (12.11) for three different tasks, we would get three different sets of w_is without any common support, that is, the positions of the non-zero values in different w_is could be different. The results would not be very interpretable, since we would not be able to ascertain what are the common sets of factors that are responsible for the output. A better approach would be to frame a joint problem such that a common set of explanatory variables emerges out of it.

$$[y_1 \mid y_2 \mid y_3] = A[w_1 \mid w_2 \mid w_3] + N \tag{12.12}$$

Here, y_1, y_2, and y_3 are the three tasks (marks in three subjects) and w_1, w_2, and w_3 are the corresponding weights. N is the noise (assumed to be i.i.d). For multi-task regression, we would want to have the ws to have a common support; that is, they should have non-zero values at the same positions. In other words, the matrix $[w_1 \mid w_2 \mid w_3]$ should be row-sparse. We already know the solution for this problem; it is $l_{2,1}$-norm minimization for matrices.

$$\min_{W} \|Y - AW\|_F^2 + \lambda \|W\|_{2,1} \tag{12.13}$$

Here, $Y = [y_1 \mid y_2 \mid y_3]$ and $W = [w_1 \mid w_2 \mid w_3]$. The l_1-norm promotes sparsity in the selection of rows/variables, and the l_2-norm makes sure that all the variables selected are used for explaining the output. By solving (12.13), one would be able to pick up the common variables that are responsible for all the outcomes.

Suggested Reading

R. Tibshirani, Regression shrinkage and selection via the lasso. *Journal of the Royal Statistical Society. Series B (Methodological)*, 267–288, 1996.

H. Zou and T. Hastie, Regularization and variable selection via the elastic net. *Journal of the Royal Statistical Society: Series B (Statistical Methodology)*, 67 (2), 301–320, 2005.

C. De Mol, E. De Vito and L. Rosasco, Elastic-net regularization in learning theory. *Journal of Complexity*, 25 (2), 201–230, 2009.

M. Yuan and Y. Lin, Model selection and estimation in regression with grouped variables. *Journal of the Royal Statistical Society: Series B (Statistical Methodology)*, 68 (1), 49–67, 2006.

J. Liu, S. Ji and J. Ye, Multi-task feature learning via efficient $l_{2,1}$-norm minimization. In *Proceedings of the Twenty-Fifth Conference on Uncertainty in Artificial Intelligence*, AUAI Press, Corvallis, OR, pp. 339–348, 2009.

13

Classification

In classification, one is interested in the category of a sample. Let us take a more specific example. We can take the example of fingerprint recognition. Here, the objective is to identify the person based on his/her fingerprint. The task of the classifier is elucidated in Figure 13.1.

A fingerprint is input to the classifier, and the classifier tells us the identity of the person based on the input. However, the classifier cannot work without any prior data. It needs fingerprints in its database from these people in order to match a test fingerprint. These samples are called training data.

Sometimes, for matching, the fingerprints are just stored; the classifier doesn't do anything else until a new fingerprint is there, and it is asked to match. Such classifiers are "Lazy." There is a more proactive version of classifiers. They learn a "model" from the training samples. They do not need to store the training samples anymore. The learnt model is capable to identify the test sample. This phase, when the classifier learns the model, is called "Training."

Classification problems arise in a myriad of areas—iris recognition, gait recognition, face recognition, speech recognition, character recognition, etc. Classification is used extensively in medicine—electroencephalography (EEG) signals are used to routinely identify stroke and Alzheimer's, and sometimes, they are even used for predicting the subject's thought in brain–computer interfaces. Classification is also used profusely in medical imaging—the task may be to judge the condition of a particular tissue, if it is malignant or benign or what is the degree of malignancy. Various problems in remote sensing are cast as classification problems as well, for example, detecting mines or segregating land mass into categories such as urban, farmland, and forest.

There are many other examples of classification problems in data analytics, for example, understanding if an email is a spam or not. Various computer vision tasks such as character recognition and object recognition are also basically classification tasks.

There are various time-tested approaches to classification; you will always hear about support vector machine (SVM), artificial neural network (ANN), random forest, and much simpler k-nearest neighbors (KNN). There are many others, but these are the ones you will hear the most about. The classifier

FIGURE 13.1
Classification.

of interest is called the sparse representation-based classification (SRC). It is relatively new, less than 10 years old. However, it is already a major player in the classification arena, owing to its simplicity and effectiveness.

Sparse Representation-Based Classification

SRC assumes that the training samples of a particular class approximately form a linear basis for a new test sample belonging to the same class. We can write the aforesaid assumption formally. If v_{test} is the test sample belonging to the kth class, then,

$$v_{test} = \alpha_{k,1}v_{k,1} + \alpha_{k,2}v_{k,2} + \ldots + \alpha_{k,n_k}v_{k,n_k} + \varepsilon \tag{13.1}$$

where $v_{k,i}$ denotes the ith training sample of the kth class and ε is the approximation error.

In a classification problem, the training samples and their class labels are provided. The task is to assign the given test sample with the correct class label. This requires finding the coefficients $\alpha_{k,i}$ in equation (13.1). Since the correct class is not known, SRC represents the test sample as a linear combination of all training samples from all classes,

$$v_{test} = V\alpha + \varepsilon \tag{13.2}$$

$$\text{where } V = \left[\underbrace{v_{1,1}|\ldots|v_{1,n_1}}_{v_1} \underbrace{v_{2,1}|\ldots|v_{2,n_2}}_{v_2} \ldots \underbrace{v_{C,1}|\ldots|v_{C,n_C}}_{v_C} \right] \text{ and}$$

$$\alpha = \left[\underbrace{\alpha_{1,1},\ldots,\alpha_{1,n_1}}_{\alpha_1} \underbrace{\alpha_{2,1},\ldots,\alpha_{2,n_2}}_{\alpha_2} \ldots \underbrace{\alpha_{C,1},\ldots,\alpha_{C,n_C}}_{\alpha_C} \right]^T$$

According to the SRC assumption, only the training samples from the correct class should form a basis for representing the test sample—samples from other classes should not. Based on this assumption, it is likely that the vector α is sparse; it should have non-zero values corresponding to the correct class and zero values for other classes. Thus (13.2) is a linear inverse problem with a sparse solution. For finding the coefficient α, a sparsity-promoting l_1-norm minimization is solved,

$$\min_{\alpha} \|v_{test} - V\alpha\|_2^2 + \lambda \|\alpha\|_1 \tag{13.3}$$

Once a sparse solution of α is obtained, the following classification algorithm is used to determine the class of the test sample:

1. Solve the optimization problem expressed in (13.1).
2. For each class k, repeat the following two steps:
   a. Reconstruct a sample for each class by a linear combination of the training samples belonging to that class by the equation $v_{recon}(k) = V_k\alpha_k$.
   b. Find the error between the reconstructed sample and the given test sample by, $error(v_{test}, k) = \|v_{test} - v_{recon}(k)\|_2$
3. Once the error for every class is obtained, assign the test sample to the class having the minimum error.

Here, the assumption is that the representative sample for the correct class will be similar to the test sample. There can be other ways for assigning the class. For example, classification results may be obtained just by looking at the magnitude of the coefficient α. Assuming that the SRC assumption holds true, values in α corresponding to the correct class should only be non-zero, and values for other classes should be zero or close to zero. Therefore, one can assign the test sample to the class having the highest norm $\|\alpha_k\|$.

The main workhorse behind the SRC is the l_1-norm minimization (13.3).

The diagrammatic representation of the inverse problem with a sparse solution is shown in Figure 13.2. The forward problem (assumption) is such

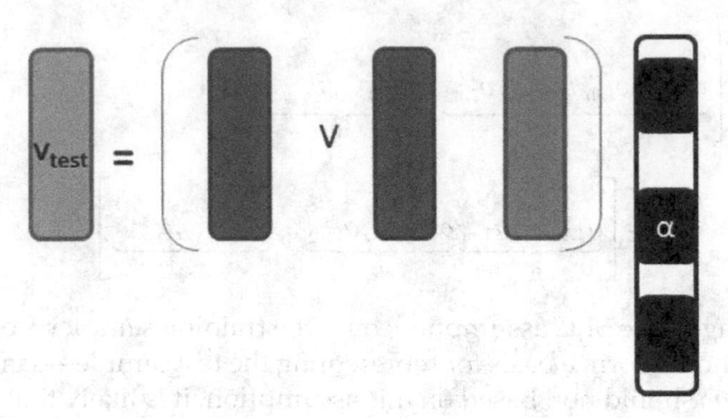

FIGURE 13.2
Sparse inverse problem.

that the vector v_{test} is formed by a linear combination of a few columns (training samples of correct class) of V. In the inverse problem, given v_{test} (test sample), few columns of V that are used to represent the test sample are selected and the corresponding sparse coefficient (α) is found.

In Figure 13.1, the colored blocks in V represent a subspace. Therefore, v_{test} can be modeled as a union of subspaces. Analyzing l_1 minimization as finding the union of subspaces is a powerful tool for studying the success of the SRC. The SRC has been very successful in face recognition problems. In a simplistic view, one can assume that each person's face can be represented roughly by three subspaces—frontal, left profile, and right profile. A test (face) sample falls in either one of these subspaces. The l_1-norm minimization basically selects the corresponding subspace and the corresponding coefficients. For the correct subspace, the residual error (ε) is small. Therefore, assigning the test sample based on small residual error is a sound criterion.

Elastic Net

There is a limitation to the l_1-norm regularization for solving the sparse recovery problem (13.3). If there is a group of samples whose pair-wise correlations are very high, then l_1 minimization tends to select one sample only from the group.

In a classification problem, training samples belonging to the same class are correlated with each other. In such a situation, the l_1 minimization used by the SRC tends to select only a single training sample from the entire class. Thus, in the extreme case, the SRC becomes a scaled version of the nearest neighbors (NN) classifier.

For explaining this effect of the least angle selection and shrinkage operator (LASSO) regularization, we rewrite the assumption expressed in equation (13.1),

$$v_{test} = \alpha_{k,1}v_{k,1} + \alpha_{k,2}v_{k,2} + \ldots + \alpha_{k,n_k}v_{k,n_k} + \varepsilon$$

where the v_ks belong to the same class and are correlated with each other. If algorithm 1, given after equation (13.3), is employed for classifying the test sample, then (in the extreme case), we find that:

1. The LASSO regularization tends to select only one of the training samples from the group. We call it $v_{i,best}$.
2. Step 2 is repeated for each class.
   a. The reconstructed vector becomes a scaled version of the selected sample, that is, $v_{recon}(i) = \alpha_{i,best}v_{i,best}$.
   b. The error from the reconstructed vector is calculated, that is, $error(v_{test}, i) = \|v_{k,test} - \alpha_{i,best}v_{i,best}\|_2$.
3. The class with the minimum error is assumed to be the class of the test sample.

The minimum LASSO error in Step 2b is $\| v_{k,test} - \alpha_{k,best}v_{k,best} \|_2$. In NN classification, the criterion for choosing the class of the test sample is $\| v_{k,test} - v_{i,j} \|_2 \; \forall j \in$ class i. This error is minimized when $v_{i,j} = v_{k,best}^{NN}$ and is given by $\|v_{k,test} - v_{k,best}^{NN}\|_2$. The LASSO error and the NN error are the same except for the scaling factor ($\alpha_{k,best}$).

When the training samples are highly correlated (which generally is the case in classification), employing l_1-norm regularization forms a serious limitation to the sparse classification problem. Decision regarding the correct class of the test sample should depend on all the training samples belonging to a class. But l_1-norm favors selecting a single training sample. This problem needs to be alleviated.

The problem of selecting a sparse group has been extensively studied in an alternate regularization called "Elastic Net" that promotes the selection of sparse groups. We apply this regularization to the classification problem.

We repeat the optimization problem used in SRC (13.3),

$$\min_{\alpha} \|v_{k,test} - V\alpha\|_2 + \lambda\|\alpha\|_1 \tag{13.4}$$

To promote selection of correlated samples, elastic net regularization proposes the following optimization problem,

$$\min_{\alpha} \|v_{k,test} - V\alpha\|_2 + \lambda_1\|\alpha\|_2^2 + \lambda_2\|\alpha\|_1 \tag{13.5}$$

The l_1-norm penalty in the above expression promotes sparsity of the coefficient vector α, while the quadratic l_2-norm penalty encourages grouping effect, that is, selection of correlated training samples. The combined effect of the mixed penalty term is that it enforces group sparsity, that is, the recovery of one or very few groups of correlated samples.

The classification is performed by algorithm 1, but instead of solving the optimization problem in equation (13.4), we need to solve the problem in equation (13.6).

Block/Joint Sparse Classification

The SRC employs an l_1 minimization for solving the inverse problem. This is an unsupervised approach; it does not utilize information about the class labels. Some argued that α is supposed to be non-zero for all training samples corresponding to the correct class. The SRC assumes that the training samples for the correct class will be automatically selected by imposing the sparsity-inducing l_1-norm; it does not explicitly impose the constraint that, if one class is selected, all the training samples corresponding to that class should have corresponding non-zero values in α. The hope is that better classification can be obtained if selection of all the training samples within the class is enforced. This was achieved by employing a supervised $l_{2,1}$-norm instead of the l_1-norm.

$$\min_{\alpha} \left\| v_{test} - V\alpha \right\|_2^2 + \lambda \left\| \alpha \right\|_{2,1} \tag{13.6}$$

where the mixed norm is defined as $\left\| \alpha \right\|_{2,1} = \sum_{k=1}^{C} \left\| \alpha_k \right\|_2$.

The inner l_2-norm enforces selection of all the training samples within the class, but the sum of l_2-norm over the classes acts as an l_1-norm over the selection of classes and selects very few classes. The block sparsity promoting $l_{2,1}$-norm ensures that if a class is selected, ALL the training samples within the class are used to represent the test sample.

The block sparse classification (BSC) approach sounds effective for general-purpose classification problems and was shown to perform well for simple classification problems. But it yields worse results than the SRC for face recognition.

To analyze this phenomenon, we refer to Figure 13.2. In BSC, all the training samples from the same class have the same class label. Therefore, the $l_{2,1}$ minimization tries to select all the training samples to represent the test sample. It treats all the colored blocks in Figure 13.2 as a single subspace

FIGURE 13.3
Frontal view and profile view.

instead of a union of subspaces; this may not be the right approach in all situations. Enforcing block sparsity is a good idea when the classification problem is simple and all the samples truly belong to a single subspace, for example, in fingerprint recognition and in character recognition. It prevents selection of samples from arbitrary classes. However, face recognition does not satisfy this simplistic assumption. Face images can belong to multiple subspaces, for example, see Figure 13.3—frontal and profile. The BSC tries to combine all the subspaces into a single one, for example, if the test sample is a left profile, it will try to fit the left and right profiles as well as the frontal view to the test sample. This is clearly error prone. Hence, it is not surprising that the BSC fails for face recognition problems, especially in challenging situations where there is a large variability in the training and test samples.

Non-linear Extensions

There are several non-linear extensions to the SRC approach. The linearity assumption in SRC can be generalized to include non-linear (polynomial) combinations. The generalization of (13.2) leads to:

$$v_{test} = f(V\alpha) + \varepsilon \tag{13.7}$$

The assumption is that the test sample can be represented as a non-linear combination of the training samples. Note that this is different from the kernel-based techniques. In these studies, the recovery of the coefficient vector required solving a non-linear inverse problem with sparsity constraints,

$$\min_{\alpha} \left\| v_{test} - f(V\alpha) \right\|_2^2 + \lambda \left\| \alpha \right\|_1 \tag{13.8}$$

There are no off-the-shelf solutions to problem (13.8). One way to solve these problems is via FOCally Underdetermined System Solver (FOCUSS) or non-linear orthogonal matching pursuit (OMP). The non-linear extension showed decent results on generic classification problems.

Several studies independently proposed the kernel sparse representation-based classification (KSRC) approach. KSRC is a simple extension of the SRC, using the kernel trick. The assumption here is that the non-linear function of the test sample can be represented as a linear combination of the non-linear functions of the training samples, that is,

$$\phi(v_{test}) = \phi(V)\alpha + \varepsilon \tag{13.9}$$

Here, $\phi(.)$ represents a non-linear function (unknown).

The simplest way to apply the kernel trick is to pre-multiply by (13.9) $\phi(V)^T$. This leads to,

$$\phi(V)_T \phi(v_{test}) = \phi(V)^T \phi(V)\alpha + \varepsilon \tag{13.10}$$

Equation (13.10) consists of inner products between the training samples and the test sample on the left-hand side and inner products between the training samples on the right-hand side. Once we have the representation in terms of inner products, applying the kernel trick is straightforward. The kernel is defined as,

$$\kappa(x_i, x_j) = \langle \phi(x_i), \phi(x_j) \rangle \tag{13.11}$$

where $\langle .,. \rangle$ represents the inner product.

Applying the kernel trick allows to represent (13.10) in the following form,

$$v_{test}^{\kappa} = V^{\kappa}\alpha + \varepsilon \tag{13.12}$$

Here, the superscript κ represents the kernelized version of the test sample and training data. It is easy to solve (13.12) by using any standard l_1 solver.

Extension to Multiple Measurement Vectors

So far, we have discussed scenarios where the objective is to classify a single instance of the test sample. This is dubbed as the single measurement vector (SMV) problem. In many scenarios, the test sample is naturally composed of multiple samples, for example, videos or multi-spectral imaging. These are called the multiple measurement vector (MMV) problems.

Assume the problem of video-based face recognition. Here, the training data V consists of video frames from person. There are multiple training samples in each class; the structure of V is therefore the same as in previous (SMV) problems. But the test sample consists of a video sequence containing multiple frames (assuming T frames), then $V_{test} = \left[v_{test}^1 \middle| \ldots v_{test}^T \right]$. One assumes that for each frame, the SRC assumption holds, that is,

$$v_{test}^i = V\alpha^{(i)} + \varepsilon, \forall i \in \{1 \ldots \mathrm{T}\} \tag{13.13}$$

This can be combined for all the frames in the following fashion,

$$V_{test} = VZ + \varepsilon \tag{13.14}$$

where $Z = [\alpha^{(1)} \mid \ldots \mid \alpha^{(T)}]$.

According to the SRC assumption, each of the $\alpha^{(i)}$'s will be sparse. The non-zero values will correspond to training samples of the correct class. Every $\alpha^{(i)}$ is supposed to be represented by training samples of the correct class. Therefore, the sparsity signature (positions of non-zero values) is expected to remain the same in all $\alpha^{(i)}$'s. If this assumption holds, Z is supposed to be row-sparse, and only those rows that correspond to the correct class will have non-zero values.

The row-sparsity assumption leads to solving the inverse problem (13.14) via,

$$\min_{Z} \left\| V_{test} - VZ \right\|_F^2 + \lambda \left\| Z \right\|_{2,1} \tag{13.15}$$

where $\left\| Z \right\|_{2,1} = \sum_j \left\| Z^{j\rightarrow} \right\|_2^2$ and $Z^{j\rightarrow}$ represents the jth row of Z.

The argument for using the mixed $l_{2,1}$-norm is the same as before. The l_2-norm over the rows promotes a dense (non-zero) solution within the selected row, but the outer l_1-norm enforces sparsity on the selection of rows. It should be noted that the $l_{2,1}$-norm in the current case is defined on a matrix; it should not be confused with the BSC assumption of block sparsity, where it was defined on a vector.

The final classification decision is similar to the SRC approach. Once Z is recovered, the class representation can be obtained by partitioning Z according to the classes. The MMV test sample V_{test} can be assigned to the class having the minimum residual error.

Group Sparse Representation-Based Classification

Consider the problem of multimodal biometrics. For each person, we have data regarding several modalities such as fingerprint, iris, and face. The task is to identify the person based on multiple sources of data. In this section, we will refer to this problem for all our discussions; however, the techniques developed herein are general enough to be applied to other problems as well.

For acquiring training data in multimodal biometrics, different biometric traits of each person are obtained; for example, fingerprint, face images, iris, and palm print may be acquired for each person. During testing, the same biometric traits are acquired for the subject, and these traits are matched with the training data for classification of the subject.

Here (in general), we assume that there are N biometric modalities. For each modality, we assume that the SRC model holds; that is, the test sample from that modality can be expressed as a linear combination of the training samples of the correct class from the same modality,

$$v_{test}^{i} = V^{i}\alpha^{i} + \varepsilon, \forall i \in \{1 \ldots N\} \quad (13.16)$$

It is possible to solve each of modalities by using the SRC approach and combine them at a later stage by using some arbitrary fusion rule. But such an approach is not elegant and does not exploit the intrinsic structure of the problem.

A better approach is to combine all the modalities into a single framework. This can be succinctly represented as,

$$\begin{bmatrix} v_{test}^{1} \\ \cdots \\ v_{test}^{N} \end{bmatrix} = \begin{bmatrix} V^{1} & \cdots & 0 \\ \cdots & \cdots & \cdots \\ 0 & \cdots & V^{N} \end{bmatrix} \begin{bmatrix} \alpha^{1} \\ \cdots \\ \alpha^{N} \end{bmatrix} + \varepsilon \quad (13.17)$$

Since each of the α^{i}'s are sparse, the simplest way to solve (13.17) would be impose a sparsity penalty and solve it via l_1 minimization. However, such a naïve approach would neither be sub-optimal, nor exploit the structure of problem.

The coefficient vector (for simplicity shown as a row vector) for each modality can be expanded as, $\alpha^i = [\alpha_1^i,\ldots,\alpha_k^i,\ldots,\alpha_C^i]$ where α_k^i denotes the coefficients corresponding to the kth class for the ith modality.

Assuming that the test sample belongs to the kth class, coefficients from this class will be non-zero.

Consider all the modalities,

$$\alpha = \left[\underbrace{\alpha_1^1,\ldots,\alpha_k^1,\ldots\alpha_C^1}_{\alpha^1} \ldots \underbrace{\alpha_1^i,\ldots,\alpha_k^i,\ldots\alpha_C^i}_{\alpha^i} \ldots \underbrace{\alpha_1^N,\ldots,\alpha_k^N,\ldots\alpha_C^N}_{\alpha^N} \right]$$

When the SRC assumptions holds true for individual modalities, the α_k^i's for each i (modality) will have non-zero values. There α will have a group sparse structure, where the non-zero elements will occur corresponding to the indices of the kth class. This leads to a group sparse representation, where the grouping is simply based on the indices. Therefore, (13.17) can be solved using group sparsity promoting $l_{2,1}$-norm,

$$\min_{\alpha} \left\| v_{test} - V\alpha \right\|_2^2 + \lambda \left\| \alpha \right\|_{2,1} \tag{13.18}$$

where $v_{test} = \begin{bmatrix} v_{test}^1 \\ \cdots \\ v_{test}^N \end{bmatrix}$, $V = \begin{bmatrix} V^1 & \cdots & 0 \\ \cdots & \cdots & \cdots \\ 0 & \cdots & V^N \end{bmatrix}$ and $\alpha \begin{bmatrix} \alpha^1 \\ \cdots \\ \alpha^N \end{bmatrix}$.

This (group sparse representation-based classification or GSRC) formulation does not suffer from the same stringent restrictions that fraught BSC. Here, one is not trying to fit one vector (test sample) to all the subspaces simultaneously (as is done by the BSC); the test samples from each modality are fitted into the subspaces spanned by the training samples of the same modality. The GSRC formulation keeps the flexibility of the SRC approach but improves upon SRC by exploiting the multimodal biometrics' problem structure.

The classification is based on the same principle as SRC. The representative sample for each class for all the modalities is computed as,

$$v_{rep}(k) = \begin{bmatrix} V_k^i \alpha_k^i \\ \cdots \\ V_k^N \alpha_k^N \end{bmatrix} \tag{13.19}$$

The test sample is assigned to the class having the minimum residual error between the test vector and the class representative. One can also use the sum of l_1-norm of the α_k^i s for each class and assign the test sample to the class having the maximum value. Whatever be the criterion (minimum residual or maximum coefficient), the GSRC has a simple and elegant decision rule that does not require arbitrary fusion strategies.

Label-Consistent Dictionary Learning

Dictionary learning has been a well-studied topic in both signal processing and machine learning community. While, in signal processing, the emphasis was on reconstruction algorithms, the machine learning researchers worked on supervised variants of dictionary learning. There are a few thousand papers on supervised dictionary learning in major conferences and journals. It will not be remotely possible to even mention them. Rather, we will discuss one major study that is simple to understand and largely popular in the machine learning community.

Label consistent dictionary learning comes under many variants. But the main idea is to learn a dictionary and the corresponding set of coefficients, such that the coefficients are mapped on to the targets.

Mathematically, this is represented as,

$$\min_{D,Z,M} \|X - DZ\|_F^2 + \lambda \|Q - WZ\|_F^2 \tag{13.20}$$

Here, X is the training data stacked as columns. D and Z are the corresponding dictionary and representation/coefficients. Q is the target formed by one hot encoding; that is, say, for a three-class problem, it will be

$$\begin{bmatrix}1\\0\\0\end{bmatrix} \text{ for class 1, } \begin{bmatrix}0\\1\\0\end{bmatrix} \text{ for class 2, and } \begin{bmatrix}0\\0\\1\end{bmatrix} \text{ for class 3.}$$

W is the linear map that needs to be learnt.

One can interpret it as a neural network. This is shown in the following Figures 13.4 and 13.5. The first part of the network is standard input and the hidden layer; note that, unlike a standard neural network, the connections are fed backward—this is owing to the synthesis nature of dictionary learning. The second layer maps the representation to the targets.

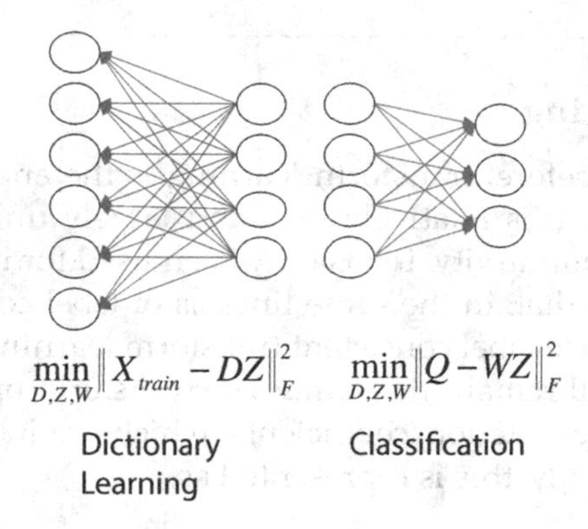

FIGURE 13.4
Label consistent dictionary learning.

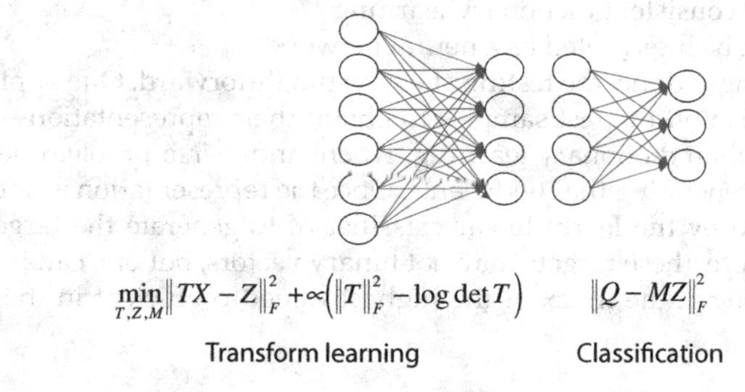

FIGURE 13.5
Label consistent transform learning.

During testing, the learnt dictionary is used to generate the coefficient. This is then mapped to the target by applying the learnt linear classifier *W*. However, it is unlikely that the resulting target vector will be binary. But one can always infer the class of the sample by finding the index of the highest-valued coefficient from the resulting target vector.

The original authors of the work added several bells and whistles to this basic formulation. This approach is also known as discriminative dictionary learning. But the basic formulation given here (and the code) yields better results than all known aforesaid formulations.

Transform Learning

As we have learnt before, transform learning is the analysis equivalent of dictionary learning. It is relatively new and largely unknown outside the signal processing community. It has only been used for inversion problems. But we have shown that, in the same lines as of label consistent dictionary learning, one can have label consistent transform learning.

The basic idea will remain the same. The transform operates on the data to produce the representation/coefficients, which are in turn mapped onto targets. Mathematically, this is represented as,

$$\min_{T,Z,M} \|TX - Z\|_F^2 + \mu\left(\|T\|_F^2 - \log\det T\right) + \lambda\|Q - MZ\|_F^2 \tag{13.21}$$

As before, the first and the second terms pertain to standard (unsupervised) transform learning. The third term is the label consistency term, as introduced in label consistent dictionary learning.

This too can be interpreted as a neural network.

Once training is done, the testing stage is straightforward. One applies the learnt transform on the test samples to obtain their representation—this is even simpler than dictionary learning, where an inverse problem needs to be solved for generating the coefficients. Once the representation is obtained, it is multiplied by the learnt linear classifier M to generate the targets. As discussed before, these targets may not binary vectors, but one can infer the class by looking at the index of the highest-valued coefficient in the target vector.

Suggested Reading

J. Wright, A. Y. Yang, A. Ganesh, S. S. Sastry and Y. Ma, Robust face recognition via sparse representation. *IEEE Transactions on Pattern Analysis and Machine Intelligence*, 31, 210–227, 2009.

A. Majumdar and R. K. Ward, Robust classifiers for data reduced via random projections, *IEEE Transactions on Systems, Man, and Cybernetics, Part B*, 40 (5), 1359–1371, 2010.

J. Yin, Z. Liua, Z. Jina and W. Yang, Kernel sparse representation based classification, *Neurocomputing*, 77 (1), 120–128, 2012.

L. Zhang, W. D. Zhou, P. C. Chang, J. Liu, Z. Yan, T. Wang, and F. Z. Li, Kernel sparse representation-based classifier, *IEEE Transactions on Signal Processing*, 60 (4), 1684–1695, 2012

Appendix: MATLAB® Codes

SRC

```
function [predClass, relative_error] =
Sparse_Classifier(trn, tst)

% --- INPUTS
% Train.X - training samples
% Train.y - training labels
% Test.X - testing samples
% Test.y - testing labels
% index - the power to which the features vectors are
to be raised for
% classification
% --- OUTPUTS
% result - recognition result for entire test data
% relative_error - relative recognition error

% Linear Classifier that solves y = Ax subject to min||x||_1
% y - normalised test sample
% A - normalised training samples
% x - sparse classification vector

Train.X = trn.X; Train.y = trn.y;
Test.X = tst.X; Test.y = tst.y;

% Normalising the training set
for i=1:size(Train.X,2)
    Atilde(:,i)=Train.X(:,i)/norm(Train.X(:,i),2);
end

k = max(Train.y); % number of classes
n_test = size(Test.X,2); % number of test data to be classified

% configuring the SPG L1 solver
options.iterations = 20;options.verbosity = 0;

% start classifcation
for iter = 1:n_test % looping through all the test samples
    ytilde = Test.X(:,iter);
    ytilde = ytilde/norm(ytilde); % normalizing the test
sample
```

```
    xp = spgl1( Atilde, ytilde, 0, 1e-1, [], options );
% calling the SPG L1 solver
    % decision making
    for i=1:k
        classRep = Atilde(:,find(Train.y==i))*xp(find(Train.
        y==i));
        residual(i) = norm(ytilde-classRep);
    end
    predClass(iter) = find(residual == min(residual));
% storing the classification results
end

% finding the relative classification error
relative_error = size(find(abs(Test.y-predClass)),2)/
size(Test.y,2);
```

GSRC

```
function [predClass, residual,xp] =
GroupSparse_Classifier2(Train, Test)

% --- INPUTS
% Train.X - training samples
% Train.y - training labels
% Test.X - testing samples
% Test.y - testing labels

% --- OUTPUTS
% result - recognition result for entire test data

% Group sparse Linear Classifier that solves y = Ax
subject to min||x||_2,1
% y - normalised test sample
% A - normalised training samples
% x - sparse classification vector


% Normalising the training sets
for i=1:size(Train.X,2)
    for j = 1:size(Train.X{i},2)
        A{i}(:,j)=Train.X{i}(:,j)/norm(Train.X{i}(:,j),2);
    end
end

TrMat=[];
for i=1:size(Train.X,2)
```

```
    TrMat = blkdiag(TrMat, A{i});
end
k = max(Train.y{1}); % number of classes

n_test = size(Test.X{1},2); % number of test
samples to be classified

% configuring the SPG solver
options.iterations = 20; options.verbosity = 0;
xp = cell(n_test,size(Train.X,2));
% define group structure
groups = [];
for i=1:size(Train.X,2)
    groups = [groups' 1:length(Train.y{i})]';
end

%start classification
for iter = 1:n_test
    iter% looping through all the test samples
    y = [];
    for i=1:size(Test.X,2)
        y = vertcat(y, Test.X{i}(:,iter)/norm(Test.X{i}
        (:,iter)));
    end

    coeff = spg_group(TrMat,y,groups,[],options); % calling
the group sparse SPG solver

    for i=1:size(Train.X,2)
        xp{iter,i} = coeff(length(Train.y{i})*(i-1)+1:
        i*length(Train.y{i}));
    end

    % decision making
    for j=1:k
        classRep = [];
        for i = 1:size(Train.X,2)
            tempRep =
A{i}(:,find(Train.y{i}==j))*xp{iter,i}(find(Train.y{i}==j));
            classRep = vertcat(classRep, tempRep);
        end
        residual(j,iter) = norm(y-classRep); % storing all
        the residuals
    end
    [temp, ind] = sort(residual, 'ascend');
    predClass(iter) = ind(1); % storing the classification
results
    end
```

Face Recognition Demo with Yale Database

```
% Face Recognition Demo
% Requires STPRTOOL for PCA and LDA - otherwise needs to be
changed

load Yale % Yale Database

dim = 20; % dimensionality of the feature

NumEl = 1:length(Data.y);

% LOO Cross Validation
for i = 1:length(Data.y)
    TestSetInd = i;
    TrainSetInd = setdiff(NumEl,i);
    Train.X = Data.X(:,TrainSetInd);
    Train.y = Data.y(:,TrainSetInd);
    Test.X = Data.X(:,TestSetInd);
    Test.y = Data.y(:,TestSetInd);

    PCAmodel = pca(Train.X,dim);
    PCATrain.X = linproj(Train.X,PCAmodel);
    PCATrain.y = Train.y;
    PCATest.X = linproj(Test.X,PCAmodel);
    PCATest.y = Test.y;

    [PCAypred(i)] = Sparse_Classifier(PCATrain, PCATest);
% sparse classifier with only PCA

    trn.X{1} = PCATrain.X;
    trn.y{1} = PCATrain.y;
    tst.X{1} = PCATest.X;
    tst.y{1} = PCATest.y;

    [ypred(i)] = GroupSparse_Classifier2(trn, tst);
% group sparse classifier with only PCA

    yactual(i) = PCATest.y;

end
PCAErr = length( find (PCAypred - yactual))/length(ypred)
CombinedErr = length( find (ypred - yactual))/length(ypred)
%% 2 features
% LOO Cross Validation
```

```
for i = 1:length(Data.y)
    TestSetInd = i;
    TrainSetInd = setdiff(NumEl,i);
    Train.X = Data.X(:,TrainSetInd);
    Train.y = Data.y(:,TrainSetInd);
    Test.X = Data.X(:,TestSetInd);
    Test.y = Data.y(:,TestSetInd);

    LDAmodel = lda(Train,dim);
    LDATrain = linproj(Train,LDAmodel);
    LDATest = linproj(Test,LDAmodel);

    [LDAypred(i)] = Sparse_Classifier(LDATrain, LDATest);
% sparse classfier with only LDA

    PCAmodel = pca(Train.X,dim);
    PCATrain.X = linproj(Train.X,PCAmodel);
    PCATrain.y = Train.y;
    PCATest.X = linproj(Test.X,PCAmodel);
    PCATest.y = Test.y;

    trn.X{1} = PCATrain.X;
    trn.y{1} = PCATrain.y;
    tst.X{1} = PCATest.X;
    tst.y{1} = PCATest.y;

    trn.X{2} = LDATrain.X;
    trn.y{2} = LDATrain.y;
    tst.X{2} = LDATest.X;
    tst.y{2} = LDATest.y;

    [ypred(i)] = GroupSparse_Classifier2(trn, tst);
% group sparse classifier with combined PCA and LDA

    yactual(i) = LDATest.y;

end

LDAErr = length( find (LDAypred - yactual))/length(ypred)
CombinedErr = length( find (ypred - yactual))/length(ypred)
%% 3 featues

% LOO Cross Validation
for i = 1:length(Data.y)
    P = randn(40, size(Data.X,1));
```

```
    TestSetInd = i;
    TrainSetInd = setdiff(NumEl,i);
    Train.X = Data.X(:,TrainSetInd);
    Train.y = Data.y(:,TrainSetInd);
    Test.X = Data.X(:,TestSetInd);
    Test.y = Data.y(:,TestSetInd);

    RandTrain.X = P*Train.X;
    RandTrain.y = Train.y;
    RandTest.X = P*Test.X;
    RandTest.y = Test.y;

    [Randypred(i)] = Sparse_Classifier(RandTrain, RandTest);
% sparse classfier with only LDA

    PCAmodel = pca(Train.X,dim);
    PCATrain.X = linproj(Train.X,PCAmodel);
    PCATrain.y = Train.y;
    PCATest.X = linproj(Test.X,PCAmodel);
    PCATest.y = Test.y;

    trn.X{1} = PCATrain.X;
    trn.y{1} = PCATrain.y;
    tst.X{1} = PCATest.X;
    tst.y{1} = PCATest.y;

    LDAmodel = lda(Train,dim);
    LDATrain = linproj(Train,LDAmodel);
    LDATest = linproj(Test,LDAmodel);

    trn.X{2} = LDATrain.X;
    trn.y{2} = LDATrain.y;
    tst.X{2} = LDATest.X;
    tst.y{2} = LDATest.y;

    trn.X{3} = RandTrain.X;
    trn.y{3} = RandTrain.y;
    tst.X{3} = RandTest.X;
    tst.y{3} = RandTest.y;

    [ypred(i)] = GroupSparse_Classifier2(trn, tst);
% group sparse classifier with combined PCA and LDA

    yactual(i) = LDATest.y;
end
```

Label Consistent Dictionary Learning

```
RandErr = length( find (Randypred - yactual))/length(ypred)
CombinedErr = length( find (ypred - yactual))/length(ypred)

function [D, Z, W] = LC_DL(X,T, numOfAtoms,lambda)

Tol = 1e-3;
maxIter = 50;
lam=lambda;
cost = [];
if size(X,2) < numOfAtoms
    error('Number of atoms cannot be more than the number
    of samples');
end

%% Dictionary initializiation

% undercomplete dictionary : QR
[Q,R]=qr(X);
D = Q(:,1:numOfAtoms);
% intializing coefficients
 Z = mldivide(D,X);
 W = T/Z;
disp('Learning dictionary')

for iter = 1:maxIter
    Dprev = D;
    %Dictionary Update
    D = mrdivide(X,Z);
     for i = 1:numOfAtoms
        D(:,i) = D(:,i)/norm(D(:,i));
     end
    % Feature Update
     Z = inv(D'*D+lam*(W'*W))*(D'*X+lam*W'*T);

    % update W
      W = T/Z;
    if norm(D-Dprev, 'fro') < Tol
        break
    end
    cost(iter) = norm(X-D*Z)^2;
end
```

Label Consistent Transform Learning

```
function [T, Z, W] = lcTransformLearning (X, labels,
numOfAtoms, mu, lambda, eps)

% solves ||TX - Z||_Fro - mu*logdet(T) + eps*mu||T||_Fro +
lambda||Q-WZ||_Fro

% Inputs
% Xlabels     - Training Data
% labels      - Class
% numOfAtoms  - dimensionaity after Transform
% mu          - regularizer for Tranform
% lambda      - regularizer for coefficient
% eps         - regularizer for Transform
% type        - 'soft' or 'hard' update: default is 'soft'
% Output
% T           - learnt Transform
% Z           - learnt sparse coefficients
% W           - linear map
if nargin < 6
    eps = 1;
end
if nargin < 5
    lambda = 1;
end
if nargin < 4
    mu = 0.1;
end

maxIter = 10;
type = 'soft'; % default 'soft'

rng(1); % repeatable
T = randn(numOfAtoms, size(X,1));
Z = T*X;
numOfSamples = length(labels);
if min(labels) == 0
    labels = labels + 1;
end

numOfClass = max(labels);
Q = zeros(numOfClass,numOfSamples);
for i = 1:numOfSamples
    Q(labels(i),i) = 1;
end
```

```
W = Q / Z;

invL = (X*X' + mu*eps*eye(size(X,1)))^(-0.5);

for i = 1:maxIter

    % update Transform T
    [U,S,V] = svd(invL*X*Z');
    D = [diag(diag(S) + (diag(S).^2 + 2*mu).^0.5)
zeros(numOfAtoms, size(X,1)-numOfAtoms)];
    T = 0.5*V*D*U'*invL;

    % update Coefficients Z
Z = (eye(size(W,2)) + lambda*W'*W)\(T*X + lambda*W'*Q);

    % update map
    W = Q / Z;

end
```

14

Computational Imaging

Introduction

A digital camera that costs less than a $100 has a resolution of 20 megapixels. Given the low price of sensors in the optical range, most of us do not feel the necessity to apply compressed sensing for imaging. We are lucky that the optical response of silicon matches that of the human eye; therefore, whenever there is a breakthrough in miniaturizing silicon technology, the camera industry benefits without breaking a sweat. That's the reason we have seen digital cameras costing less than $100 to pack in more pixels every year, starting from around 5 megapixels until about a decade back to 20 megapixels today.

But if we consider buying a near-infrared (NIR) camera for health monitoring/surveillance, the price shoots up. A 1-megapixel camera will cost at least $2000. This is because the moment we venture out of the optical range of the human eye, more expensive sensors (not based on silicon) are needed. This shoots up the camera price. In this chapter, we will study how compressive sensing can be used to bring down the cost of such multi-spectral cameras.

Multi-spectral cameras are still relatively expensive. They only capture 5–20 bands. When there is a need to capture hyper-spectral images—more than 100 bands—the cost of acquisition runs into millions of dollars. This is because the cost of the sensors is very high in this range, and the cost of camera is directly proportional to the number of sensors. Compressed sensing can help bring down the cost of such camera.

Compressive Imaging—Single-Pixel Camera

We studied about compressed sensing in medical imaging. It was relatively straightforward, because the measurement domain and the signal domain were different; for example, in magnetic resonance imaging (MRI), the measurement is in Fourier space, whereas the signal is in spatial (pixel) domain. Therefore, sub-sampling the Fourier domain for compressive sampling measurements was straightforward. However, in digital photography, both the measurement and signal domains are the same—pixels. Therefore, one cannot sub-sample directly; reducing the number of sensors requires some effort.

One of the most well-known techniques to drastically reduce the number of sensors is the single-pixel camera (Figure 14.1). At the heart of it is a digital micromirror array (DMD), where the orientation of the mirrors can be controlled by a random number generator (RNG). If the output is 1, the corresponding mirror points toward the lens (in the direction of the photodiode [PD]), and if the output is zero, it points away.

When in use, the scene is projected on a lens, which in turn projects on the DMD. The RNG generates a random binary sequence, which controls the orientation of the mirrors. The scene is projected on the mirror array. This basically emulates a product between the scene and the random binary sequence—it is akin to a random binary projection. The binary projection is captured by the PD. This emulates one compressed sensing measurement. The initial idea for single-pixel camera was to compress images; hence, it has the transmitter (sensing the compressively sensed samples). This compressed representation was sent to the receiver at some far end, where compressed sensing reconstruction took place.

One random configuration of the micro-mirrors represents one inner product between the image (x) and the random binary vector (b_i). For getting more compressed samples, more such random configurations are generated.

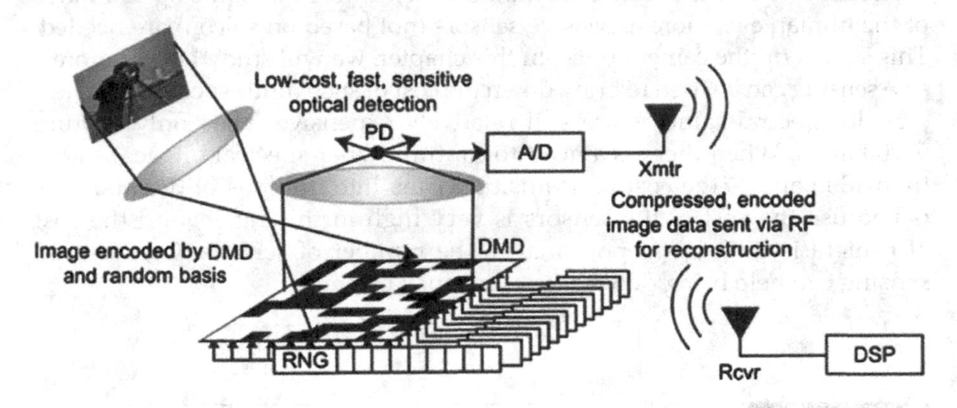

FIGURE 14.1
Single pixel camera.[1]

[1] M. F. Duarte, M. A. Davenport, D. Takhar, J. N. Laska, T. Sun, K. F. Kelly and R. G. Baraniuk, Single-pixel imaging via compressive sampling. *IEEE Signal Processing Magazine*, 25 (2), 83–91, 2008.

Generating the random configurations takes time; therefore, it is tacitly assumed that the scene (being captured) is not changing. The entire acquisition process is expressed as follows,

$$\begin{bmatrix} y_1 \\ y_2 \\ \cdots \\ y_m \end{bmatrix} = \begin{bmatrix} \vec{b}_1^T \\ \vec{b}_2^T \\ \cdots \\ \vec{b}_m^T \end{bmatrix} x \tag{14.1}$$

Here, y_1 to y_m are the measurements, b_1 to b_m are the random binary vectors (produced by random configurations of the DMD), and x is the image. This entire process emulated compressed sensing measurements. The image can be reconstruction using any standard approach.

Some reconstruction results are shown in Figure 14.2. The first one is a black-and-white image of the letter "R." The original image is of size 256 × 256 (~65000 pixels). The reconstructed image (on the right) is done from

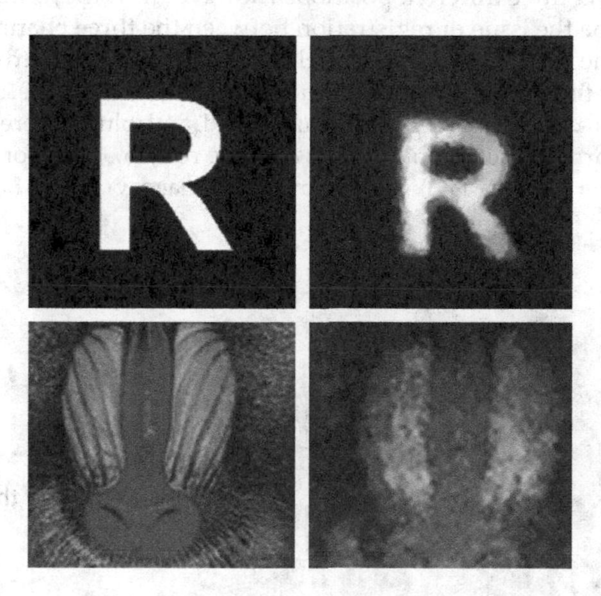

FIGURE 14.2
Sample reconstruction from single pixel camera.[2]

[2] M. F. Duarte, M. A. Davenport, D. Takhar, J. N. Laska, T. Sun, K. F. Kelly and R. G. Baraniuk, Single-pixel imaging via compressive sampling. *IEEE Signal Processing Magazine*, 25 (2), 83–91, 2008.

only 1300 binary measurements (2%) by the single-pixel camera. The second image is that of "mandrill." The original image has the same resolution, that is, 256 × 256. It has been reconstructed on the right from 10% (6500) measurements. For the mandrill image, each of the channels (red, green, and blue) is separately acquired by the single-pixel camera.

Compressive Color Imaging

The original single-pixel camera was built for grayscale imaging. Later, it was modified for color imaging. In regular color imaging, one uses the single-sensor architecture (different from single-pixel camera). As we all know, color images are captured in three channels—red, green, and blue; one ideally needs three different sensors for each channel. But three sensors for each pixel would pose difficulties. First, since the sensors will be physically located at three different positions (however near they might be), there will always be the issue of registration between the three channels. Second, the cost of one sensor for each channel would increase the hardware cost.

The Bayer filter addresses this issue. It is designed on the fact that our eyes are more sensitive to green than to red and blue. Therefore, 50% of the sensors are for the green channel and the rest 25% each for the red and blue channels (Figure 14.3). Capturing the image via this Bayer filter is

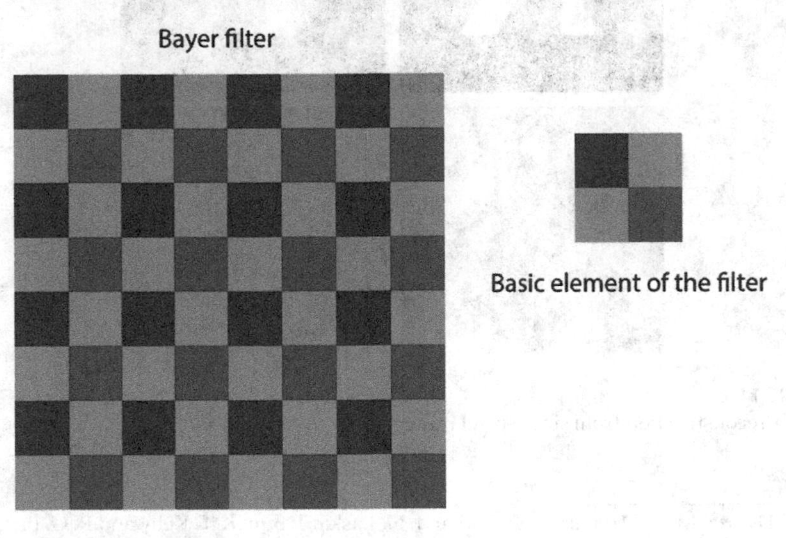

FIGURE 14.3
Bayer filter.

called mosaicing. From the mosaiced images, one needs to recover each of the three channels by some kind of interpolation—a technique known as demosaicing.

While modifying the single-pixel camera for color imaging, the Bayer filter was incorporated. The Bayer pattern is imprinted in the DMD. The random number generator separately controls the red, green, and blue mirrors. When the scene is projected, there is a rotating color filter that projects through either of the red, green, and blue filters. When the scene is being filtered through the blue channel, only the corresponding (blue) mirrors of the DMD are flipped. The same happens for the other two channels.

Effectively, this process acquires an image that is a projection of the single-sensor image acquired via Bayer filter of a color camera. Therefore, compressed sensing reconstruction produces a mosaiced image. One needs to demosaic the image to obtain the full-color image.

Some results are shown in Figure 14.5. The leftmost image is the ground truth. The second and third images are from two variants of the method proposed by proponents of compressive color imaging[3]; these are acquired via the camera shown in Figure 14.4. The rightmost image (in Figure 14.5) is from compressed sensing of each of the channels separately (acquired via single-pixel camera). The acquisition is carried out at 25% under-sampling. The results show that the Bayer-incorporated architecture shows slight improvements over separate acquisition of each color channel.

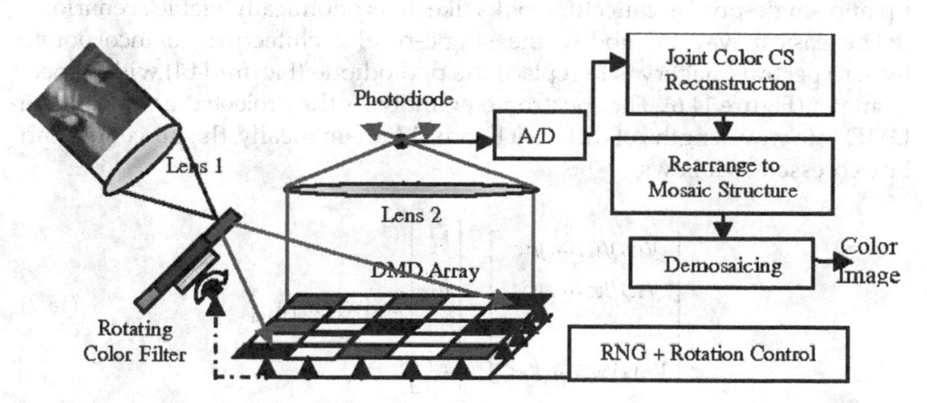

FIGURE 14.4
Compressive color imaging.[3]

[3] P. Nagesh and B. Li, Compressive imaging of color images. In *Acoustics, Speech and Signal Processing, 2009. ICASSP 2009. IEEE International Conference on* (pp. 1261–1264). IEEE, 2009.

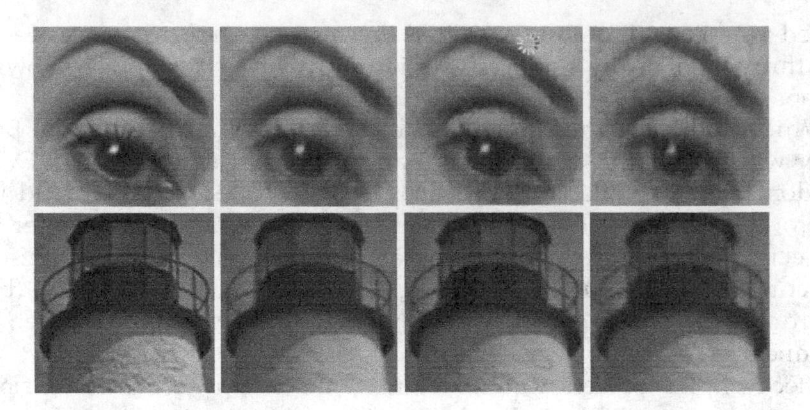

FIGURE 14.5
Results of compressive color imaging.

Compressive Hyper-spectral Imaging

As mentioned in the introduction, hyper-spectral imaging is a costly exercise. This is largely because of the cost of the sensors used for data acquisition. The application of single-pixel camera for digital photography was a toy problem, more of a proof of concept. It is never going to replace regular cameras. But once we delve outside the realm of the optical range, the cost of sensors shoots up and single-pixel architecture looks like an economically viable scenario.

The easiest way to modify the single-pixel architecture to incorporate hyper-spectral imagery is to replace the photodiode (Figure 14.1) with a spectrometer (Figure 14.6). The spectrometer analyzes the projected light from the DMD into wavelengths of different bands. Mathematically, the procedure can be expressed as follows,

$$\begin{bmatrix} y_{11}, y_{12}, \ldots, y_{1C} \\ y_{21}, y_{22}, \ldots, y_{2C} \\ \ldots \\ y_{m1}, y_{m2}, \ldots, y_{mC} \end{bmatrix} = \begin{bmatrix} \vec{b}_1^T \\ \vec{b}_2^T \\ \ldots \\ \vec{b}_m^T \end{bmatrix} [x_1 \mid x_2 \mid \ldots \mid x_C] \tag{14.2}$$

Here, we assume that there are C channels and m measurements are taken for each channel. Given the architecture, the random binary basis remains the same for each channel measurement.

There can be several ways to reconstruct the hyper-spectral datacube given the measurements. One approach is to use Kronecker compressed

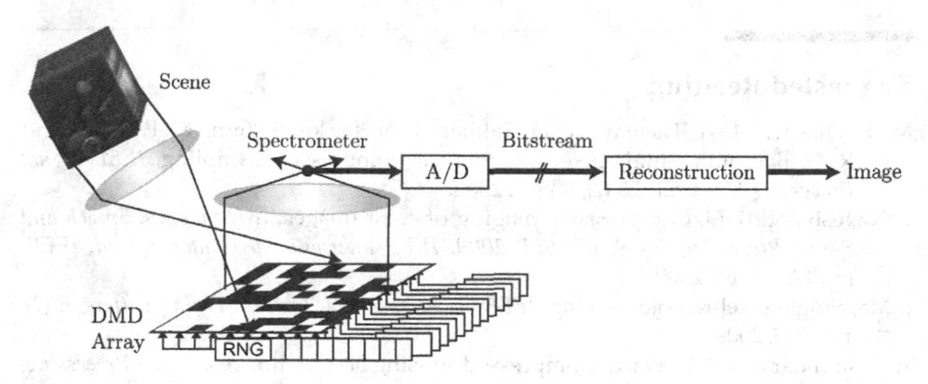

FIGURE 14.6
Single pixel architecture for hyper-spectral imaging.[4]

sensing. The spectral–spatial correlation is effectively whitened by using a three-dimensional (3D) sparsifying transform, and standard l_1 minimization is used for recovery. Another (better) approach is to exploit the spectral correlation in terms of row sparsity and only whiten the spatial correlations using a two-dimensional (2D) sparsifying transform. The $l_{2,1}$ minimization is used for recovering the datacube.

The results from two different techniques are shown in Figure 14.7. One can see that $l_{2,1}$ minimization is indeed better. It can preserve the structure and values in all locations. But the Kronecker formulation based on l_1 minimization cannot recover the values correctly.

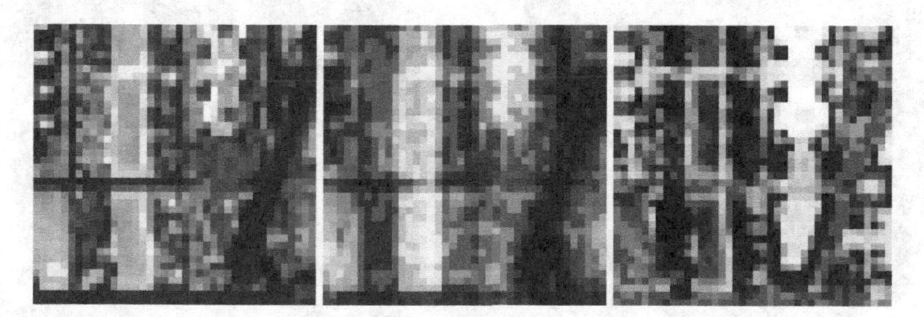

FIGURE 14.7
Left to right: original, $l_{2,1}$ minimization, and l_1 minimization.[5]

[4] https://archive.cnx.org/contents/f0bdfbd9-ec2c-40ca-bb1e-d7f025be17d9@4/hyperspectral-imaging

[5] A. Gogna, A. Shukla, H. K. Agarwal and A. Majumdar, 2014, October. Split Bregman algorithms for sparse/joint-sparse and low-rank signal recovery: Application in compressive hyperspectral imaging. In *Image Processing (ICIP), 2014 IEEE International Conference on* (pp. 1302–1306). IEEE.

Suggested Reading

M. F. Duarte, M. A. Davenport, D. Takhar, J. N. Laska, T. Sun, K. F. Kelly and R. G. Baraniuk, Single-pixel imaging via compressive sampling. *IEEE Signal Processing Magazine*, 25 (2), 83–91, 2008.

P. Nagesh and B. Li, Compressive imaging of color images. In *Acoustics, Speech and Signal Processing, 2009. ICASSP 2009. IEEE International Conference on*, IEEE, pp. 1261–1264, 2009.

J. Ma, Single-pixel remote sensing. *IEEE Geoscience and Remote Sensing Letters*, 6 (2), 199–203, 2009.

A. Majumdar and R. K. Ward, Compressed sensing of color images. *Signal Processing*, 90 (12), 3122–3127, 2010.

15

Denoising

Introduction

Signal denoising is one of the classical problems in the area of signal processing. Noise models can be categorized as additive and multiplicative. Most denoising problems fall under the former category, so we will concentrate more on that. In our context, "noise" means random perturbations. Any type of structured artefacts will not be considered as noise; for example, we will consider problems such as inpainting as denoising problems.

Let us understand some examples of how noise is introduced in signals. Electroencephalogram (EEG) signals typically sense activities from the brain. But involuntary muscle movements affect the EEG signals by introducing spikes. This is one kind of sparse noise. Such sparse noise also appears in images due to some fluctuations during analog to digital conversion. Since the fluctuations can be both positive and negative, it is typically called salt-and-pepper noise; this kind of noise is sparse as well. In magnetic resonance imaging, thermal noise is present. Unlike salt-and-pepper noise, which affects few pixels in the image severely, thermal noise affects every pixel but by a small amount. This is typically modeled as Gaussian distribution; it is the most common form of noise.

So far, we have discussed examples of additive noise. Multiplicative noise commonly appears in ultrasonography and synthetic aperture radar imaging. It is called "speckle" noise. It is a signal-dependent noise that appears due to multiple reflections from the surroundings.

There are classical techniques based on filtering for removing all kinds of signal noise. However, since the concentration of this book is on sparsity-based techniques, we will discuss only these. For classical techniques, one can read any textbook on signal and image processing.

We will start with Gaussian denoising, since it is the most popular. We will then learn about removing impulse noise. Removing speckle noise is very domain-specific; we will not discuss it in this chapter.

Gaussian Denoising

Synthesis Formulation

Let us start with Gaussian denoising, because it is the most prevalent noise and is also the most widely studied. It is an additive noise and can be modeled as,

$$y = x + n \tag{15.1}$$

The basic assumption behind sparsity-based denoising is that the signal can be sparsely represented in a transform domain but the noise cannot. If this assumption holds, one can always find a threshold (in the transform domain) that suppresses the dense but small noise and keeps the sparse high-valued signal components. After noise suppression, the denoised signal is obtained by applying the inverse transform.

Procedurally, the first step is to apply a sparsifying transform Φ (wavelet, discrete cosine transform (DCT), etc.) to (15.1). This is represented as,

$$\Phi y = \Phi x + \Phi n \tag{15.2}$$

It is supposed that the coefficients for the true signal, that is, $\alpha = \Phi x$, will be sparse, and the noise after transform $n' = \Phi n$ will remain dense (follows from the property of Gaussian). Therefore, the obtained signal $\beta = \Phi y$ will be a combination of true sparse coefficients and dense small noise. The noise can be removed by framing denoising as,

$$\min_{\alpha} \|\beta - \alpha\|_2^2 + \lambda \|\alpha\|_1 \tag{15.3}$$

We have learnt in Chapter 3 that this has a closed-form solution in the form of soft thresholding.

$$\alpha = signum(b)\max\left(0, |b| - \frac{\lambda}{2}\right) \tag{15.4}$$

This is the thresholding step, with threshold $\lambda/2$. The exact threshold value can be found out, given the statistical parameters of the noise. Once the sparse coefficients corresponding to the clean image are obtained, the image itself is recovered by applying the inverse transform.

Analysis Formulation

This is a generic approach toward denoising. But for images, one can have stronger priors such as total variation; such priors are not orthogonal or tight frames, and hence, the synthesis formulation cannot be used. In such cases, one needs the analysis formulation. This is given by,

$$\min_{x} \|y - x\|_2^2 + 2\lambda \|\Psi x\|_1 \tag{15.5}$$

We are showing this formulation for the generic case; total variation is a special case, where Ψ is vertical and horizontal differencing operator.

To minimize (15.5), we take the gradient of (15.5),

$$y - x + \lambda \Psi^T D \Psi x, \text{ where } D = diag(|\Psi x|^{-1}) \tag{15.6}$$

Setting the gradient to zero, one gets,

$$(I + \lambda \Psi^T D \Psi) x = y \tag{15.7}$$

Using the following matrix inversion lemma,

$$(I + \lambda \Psi^T D \Psi)^{-1} = I - \Psi^T \left(\frac{1}{\lambda} D^{-1} + \Psi^T \Psi \right)^{-1} \Psi \tag{15.8}$$

we arrive at the following identity,

$$x = y - \Psi^T \left(\frac{1}{\lambda} D^{-1} + \Psi^T \Psi \right)^{-1} \Psi y \tag{15.9}$$

or equivalently,

$$z = \left(\frac{1}{\lambda} D^{-1} + \Psi^T \Psi \right)^{-1} \Psi y, x = y - \Psi^T z \tag{15.10}$$

Solving z requires solving the following,

$$\left(\frac{1}{\lambda} D^{-1} + \Psi^T \Psi \right) z = \Psi y \tag{15.11}$$

Adding cz to both sides of (15.11) and subtracting $\Psi^T\Psi z$ gives,

$$z = \left(\frac{2a}{\lambda}D^{-1} + cI\right)^{-1}(cz_{k-1} + S(b - S^T z_{k-1})) \tag{15.12}$$

where c is the maximum eigenvalue of $\Psi^T\Psi$.

This gives us the update steps for analysis prior denoising. Note that we have derived the traditional algorithm for this task. One can alternately employ the more modern approach of variable splitting and alternating direction method of multipliers to arrive at a different algorithm.

Dictionary Learning

These are the standard sparsity-based approaches toward signal denoising. Here, it is assumed that the sparsity basis is known, for example, wavelet for synthesis prior or finite differencing for analysis prior. Such pre-defined basis is good for sparsifying a wide class of signals. But it is well known that such fixed basis is not the best for sparsifying a particular signal.

Let us understand the repercussions of this fact. The denoising capability is solely dependent on the sparsifying capacity of the transform. If the signal is not sparsely represented, that is, if there are components that are dense, these will be treated as noise by the thresholding operator and removed. The recovered signal will have missing components to it. This is unacceptable.

Since such fixed basis is not the answer, one needs to resort to adaptively learning the basis from the signal itself. As we have studied before, this is achievable by dictionary learning. The overall approach is to take portions of the signal and learn a basis, such that it represents the signal in a very sparse fashion.

The noise model is given by (15.1). Given an estimate of the signal $\hat{x}$, the dictionary is learnt from patches of the signal. This is expressed as,

$$P_i\hat{x} = Dz_i \ \forall i \tag{15.13}$$

Here, P_i is the patch extraction operator, D is the learnt sparsifying basis, and x_i is the corresponding sparse coefficient.

What dictionary learning does is that it incorporates a learning step into the standard synthesis prior formulation. Mathematically, this is expressed as,

$$\min_{\hat{x},D,Z} \underbrace{\|y - \hat{x}\|_2^2}_{\text{Data Fidelity}} + \mu \underbrace{\left(\sum_i \|P_i\hat{x} - Dz_i\|_2^2 + \lambda\|z_i\|_1\right)}_{\text{Dictionary Learning}} \tag{15.14}$$

The first term (data fidelity) arises from the requirement of Gaussian denoising. The dictionary learning term replaces the fixed sparsity-promoting term in the synthesis formulation.

The formulation (15.14) can be solved using the alternating direction method of multipliers, where each of the variables is updated separately.

$$\hat{x} \leftarrow \min_{\hat{x}} \|y - \hat{x}\|_2^2 + \mu \sum_i \|P_i\hat{x} - Dz_i\|_2^2 \tag{15.15}$$

$$\min_D \sum_i \|P_i\hat{x} - Dz_i\|_2^2 \equiv \|X - DZ\|_F^2 \tag{15.16}$$

$$\min_Z \sum_i \|P_i\hat{x} - Dz_i\|_F^2 + \lambda \|z_i\|_1 \equiv \|X - DZ\|_F^2 + \lambda \|Z\|_1 \tag{15.17}$$

In (15.16) and (15.17), X is formed by stacking the patches $P_i\hat{x}$ as columns of X, and Z is formed by stacking the z_is as columns.

The first two sub-problems are standard least squares, having a closed-form solution. The last one is an l_1 minimization problem, to which we know the solution.

Transform Learning

This is the synthesis prior solution to dictionary learning–based denoising. As in sparsity-based solutions, we can alternately have an analysis prior formulation to learning-based denoising. It is based on the analysis equivalent of dictionary learning and is called transform learning. The basic idea still remains the same; that is, a transform is learnt such that it represents portions of the signal in a sparse fashion. But instead of regenerating the signal from the dictionary and the coefficients, transform learning operates on the signal to produce the coefficients.

The general formulation is given by,

$$\min_{\hat{x},T,Z} \underbrace{\|y - \hat{x}\|_2^2}_{\text{Data Fidelity}} + \mu \underbrace{\left(\sum_i \|TP_i\hat{x} - z_i\|_F^2 + \lambda \|z_i\|_1 \right)}_{\text{Transform Learning}} \tag{15.18}$$

One notices that the first term remains as before—this is the data fidelity term. The term within (.) is the transform learning formulation. The solution to (15.18) proceeds in two steps. At first, the signal is updated by solving,

$$\hat{x} \leftarrow \min_{\hat{x}} \|y - \hat{x}\|_2^2 + \lambda \sum_i \|TP_i\hat{x} - z_i\|_F^2 \tag{15.19}$$

This is a simple least square update having a closed-form solution. The second step is transform learning, given by,

$$\min_{T,Z} \sum_i \|TP_i\hat{x} - z_i\|_F^2 + \lambda \|z_i\|_1 \tag{15.20}$$

This is the transform learning term. We have learnt in a previous chapter the updates for the same.

Impulse Denoising

Gaussian noise has been widely studied and is the most popular noise model. However, as mentioned earlier, there are other kinds of noise that are of high value but affect only a portion of the signal. This is impulse noise. In some applied areas of signal processing, when referring to "noise," one means Gaussian noise; impulse noise is dubbed as perturbation/artifacts.

The noise model is still additive (15.1), but the noise n is not dense; it is sparse and of high value. Therefore, instead of the Euclidean data fidelity term, one needs to employ the more robust l_1-norm data fidelity, also known as absolute distance.

Synthesis Prior

As we did for Gaussian noise, let us start with the most common approach—the synthesis prior solution to impulse denoising. Unfortunately, owing to the absolute distances, one cannot have a nice closed-form solution for the synthesis prior formulation. One needs to solve the following,

$$\min_{\alpha} \|y - \Phi^T\alpha\|_1 + \lambda \|\alpha\|_1 \tag{15.21}$$

The first term is the data fidelity term corresponding to impulse noise. Since the noise is sparse, we need to minimize its l_1-norm.

The simplest way to approach this problem is to employ the Split Bregman technique. We introduce a proxy variable $p = y - \Phi^T\alpha$. The Bregman formulation leads to the following augmented Lagrangian,

$$\min_{\alpha,p} \|p\|_1 + \lambda\|\alpha\|_1 + \mu\|p - y + \Phi^T\alpha - b\|_2^2 \tag{15.22}$$

Here, b is the Bregman relaxation term.

Using alternating direction method of multipliers, one can simplify (15.22) to the following sub-problems,

$$\alpha \leftarrow \min_{\alpha} \lambda\|\alpha\|_1 + \mu\|p - y + \Phi^T\alpha - b\|_2^2 \tag{15.23}$$

$$p \leftarrow \min_{p} \|p\|_1 + \mu\|p - y + \Phi^T\alpha - b\|_2^2 \tag{15.24}$$

The first one is a standard l_1 minimization problem. The second one has a closed for solution in the form of soft thresholding.

Finally, one needs to update the Bregman relaxation variable,

$$b \leftarrow p - y + \Phi^T\alpha - b \tag{15.25}$$

Note that in the Split Bregman formulation, one does not need to update the hyper-parameter μ. This concludes the derivation for synthesis prior impulse denoising.

Analysis Prior

In analysis prior formulation, instead of trying to recover the sparse coefficient (α), we directly retrieve the signal (x). We have learnt the formulation for analysis prior in the case of Gaussian denoising. The only difference with impulse denoising is in the change of the data fidelity term. The formulation is given as,

$$\min_{x} \|y - x\|_1 + \lambda\|\Psi x\|_1 \tag{15.26}$$

As before, we follow the Split Bregman approach to solve (15.26). We introduce two proxies,

i. $p = y - x$
ii. $z = \Psi x$

With these, the formulation (15.26) can be expressed as,

$$\min_{x,p,z} \|p\|_1 + \lambda\|z\|_1 + \mu_1\|p - y + x - b_1\|_2^2 + \mu_2\|z - \Psi x - b_2\|_2^2 \tag{15.27}$$

Here b_1 and b_2 are the Bregman relaxation variables.

Using alternating minimization, (15.27) can be segregated into the following sub-problems,

$$x \leftarrow \min_{x} \mu_1 \|p-y+x-b_1\|_2^2 + \mu_2 \|z-\Psi x-b_2\|_2^2 \quad (15.28)$$

$$p \leftarrow \min_{p} \|p\|_1 + \mu_1 \|p-y+x-b_1\|_2^2 \quad (15.29)$$

$$z \leftarrow \min_{z} \lambda \|z\|_1 + \mu_2 \|z-\Psi x-b_2\|_2^2 \quad (15.30)$$

We know the solution to each sub-problem. The first one is a simple least squares problem. The second one has a closed-form solution via soft thresholding. The third one is an l_1 minimization problem.

The final step is to update the Bregman relaxation variables.

$$b_1 \leftarrow p-y+x-b_1 \quad (15.31)$$

$$b_2 \leftarrow z-\Psi x-b_2 \quad (15.32)$$

This concludes the derivation for the analysis prior impulse denoising algorithm.

Dictionary Learning

We have studied how dictionary learning can be used for Gaussian denoising, and we have studied how the cost function changes from Gaussian to impulse denoising. By now, you would have figured out what we are going to do next. We change the cost function for impulse denoising to the following,

$$\min_{x,D,Z} \|y-x\|_1 + \mu\left(\sum_i \|P_i x - Dz_i\|_2^2 + \lambda \|z_i\|_1\right) \quad (15.33)$$

As we have been doing so far, we need to introduce a proxy variable $p = y - x$; this leads to the Split Bregman formulation,

$$\min_{x,D,Z,p} \|p\|_1 + \mu\left(\sum_i \|P_i x - Dz_i\|_2^2 + \lambda \|z_i\|_1\right) + \eta \|p-y+x-b\|_2^2 \quad (15.34)$$

Alternating minimization of (15.34) leads to the following sub-problems,

$$x \leftarrow \min_{x} \mu\left(\sum_{i}\left\|P_i x - Dz_i\right\|_2^2\right) + \eta\left\|p - y + x - b\right\|_2^2 \tag{15.35}$$

$$D,Z \leftarrow \min_{D,Z}\left(\sum_{i}\left\|P_i x - Dz_i\right\|_2^2 + \lambda\left\|z_i\right\|_1\right) \tag{15.36}$$

$$p \leftarrow \min_{p}\left\|p\right\|_1 + \eta\left\|p - y + x - b\right\|_2^2 \tag{15.37}$$

The first step is the update for the signal. This is a least squares problem having a closed-form solution. The second sub-problem is the standard dictionary learning problem. The third one has a closed-form solution in the form of soft thresholding. As before, we have to update the Bregman relaxation variable. It is a standard gradient update; therefore, we skip writing it explicitly.

Transform Learning

Although transform learning is new, it has been used for impulse denoising. The formulation is straightforward; in the standard Gaussian denoising formulation, we replace the Euclidean distance with the absolute distance. This leads to,

$$\min_{x,T,Z}\left\|y - x\right\|_1 + \mu\left(\sum_{i}\left\|TP_i x - z_i\right\|_2^2 + \lambda\left\|z_i\right\|_1\right) \tag{15.38}$$

We use the same proxy variable as dictionary learning, that is, $p = y - x$, leading to the following Split Bregman formulation,

$$\min_{x,T,Z,p}\left\|p\right\|_1 + \mu\left(\sum_{i}\left\|TP_i x - z_i\right\|_2^2 + \lambda\left\|z_i\right\|_1\right) + \eta\left\|p - y + x\right\|_2^2 \tag{15.39}$$

With alternating direction method of multipliers, we arrive at the following sub-problems,

$$x \leftarrow \min_{x} \mu\left(\sum_{i}\left\|TP_i x - z_i\right\|_2^2\right) + \eta\left\|p - y + x - b\right\|_2^2 \tag{15.40}$$

$$T,Z \leftarrow \min_{D,Z}\left(\sum_i \left\|TP_ix - z_i\right\|_2^2 + \lambda\left\|z_i\right\|_1\right) \tag{15.41}$$

$$p \leftarrow \min_p \left\|p\right\|_1 + \eta\left\|p - y + x - b\right\|_2^2 \tag{15.42}$$

The first step is the update for the signal. This is a least squares problem having a closed-form solution. The second sub-problem is the standard transform learning problem; we have studied the updates for the individual variables before, so we do not elaborate here. The third one has a closed-form solution in the form of soft thresholding. As before, we have to update the Bregman relaxation variable by a gradient step.

Suggested Reading

D. L. Donoho, I. M. Johnstone, G. Kerkyacharian and D. Picard, Wavelet shrinkage: Asymptopia?. *Journal of the Royal Statistical Society. Series B (Methodological)*, 301–369, 1995.

C. R. Vogel and M. E. Oman, Iterative methods for total variation denoising. *SIAM Journal on Scientific Computing*, 17 (1), 227–238, 1996.

M. Elad and M. Aharon, Image denoising via sparse and redundant representations over learned dictionaries. *IEEE Transactions on Image Processing*, 15 (12), 3736–3745, 2006.

Index

Note: Page numbers in italic and bold refer to figures and tables respectively.

9781032338712